"In this commanding r [illegible])reak-
ing researcher, MacArthur Foundation Fellow, founder and director of the Johns Hopkins Center for Health Equity—outlines innovative health equity solutions that can move us toward a societal 'herd immunity' where we're tackling not just clinical disease but the deep-seated impacts of structural racism."

— Garth Graham, MD, MPH, Global Head of Public Health, Google Inc. / former Deputy Assistant Secretary for Minority Health, US Department of Health and Human Services

"Dr. Cooper's personal and professional journey is both riveting and inspiring; the scenes from her childhood in Liberia alone offer a global history lesson that resonates in present-day America. The unique experiences she brings to this unprecedented moment of the intersection of community health and racial reckoning make *Why Are Health Disparities Everyone's Problem?* not only an essential read but a central question for our time."

— Marc H. Morial, President/CEO, National Urban League

"A compelling and enlightening record of Dr. Cooper's journey of awakening to the origins and widespread impacts of health disparities and to the need for health equity in local and global communities. She shares the richness of her experiences and the piercing insights that have fueled her celebrated quest to unmask the underlying causes of and to propose solutions for the pervasive and persistent disparities whose deleterious effects in disadvantaged communities have broad effects on all others."

— James R. Gavin III, MD, PhD, Emory University School of Medicine / Chairman Emeritus, Partnership for a Healthier America, and author of *Dr. Gavin's Health Guide for African Americans: How to Keep Yourself and Your Children Well*

"Drawing on a lifetime of global experiences and decades of research, Dr. Cooper convincingly argues that racial inequities are an enormous economic and moral burden that hurts all of us. With an evidence-informed optimism, *Why Are Health Disparities Everyone's Problem?* is a desperately needed, innovative playbook to tackle the unfinished chapter in America's struggle with racial inequity with renewed vigor and competence. It shows us where to begin in dismantling the upstream structural factors that create racial and socioeconomic differences in health."

— David R. Williams, MPH, PhD, Chair, Department of Social and Behavioral Sciences, Harvard T. H. Chan School of Public Health

Why Are Health Disparities Everyone's Problem?

JOHNS HOPKINS
WAVELENGTHS

In classrooms, field stations, and laboratories in Baltimore and around the world, the Bloomberg Distinguished Professors of Johns Hopkins University are opening the boundaries of our understanding on many of the world's most complex challenges. The Johns Hopkins Wavelengths series brings readers inside their stories, presenting the pioneering discoveries and innovations that benefit people in their neighborhoods and across the globe in artificial intelligence, cancer research, food systems, health equity, science diplomacy, and other critical areas of study. Through these compelling narratives, their insights will spark conversations from dorm rooms to dining rooms to boardrooms.

This print and digital media program is a partnership between the Johns Hopkins University Press and the University's Office of Research. Team members include:

Consultant Editor: Sarah Olson

Senior Acquisition Editor: Matthew R. McAdam

Copyeditor: Charles B. Dibble

Art Director: Martha Sewall

Designer: Matthew Cole

Production Supervisor: Jennifer Paulson

Publicist: Kathryn Marguy

Program Manager: Anna Marlis Burgard

JHUP Director and Publisher: Barbara Kline Pope

Office of Research Executive Director for Research: Julie Messersmith

Why Are Health Disparities Everyone's Problem?

LISA COOPER, MD, MPH

Johns Hopkins University Press
Baltimore

Printed in the United States of America on acid-free paper
9 8 7 6 5 4 3 2 1

Johns Hopkins University Press
2715 North Charles Street
Baltimore, Maryland 21218-4363
www.press.jhu.edu

Library of Congress Cataloging-in-Publication Data

Names: Cooper, Lisa A., author.
Title: Why are health disparities everyone's problem? / Lisa Cooper, MD, MPH.
Description: Baltimore : Johns Hopkins University Press, 2021. | Series: Johns Hopkins wavelengths | Includes bibliographical references and index.
Identifiers: LCCN 2020052638 | ISBN 9781421441153 (paperback ; alk. paper) | ISBN 9781421441160 (ebook) | ISBN 9781421441177
Subjects: MESH: Health Status Disparities | Vulnerable Populations | Health Equity | Healthcare Disparities | United States
Classification: LCC RA418 | NLM WA 300 AA1 | DDC 362.1--dc23
LC record available at https://lccn.loc.gov/2020052638

A catalog record for this book is available from the British Library.

Special discounts are available for bulk purchases of this book. For more information, please contact Special Sales at specialsales@jh.edu.

Johns Hopkins University Press uses environmentally friendly book materials, including recycled text paper that is composed of at least 30 percent post-consumer waste, whenever possible.

Contents

Preface

WE'RE ALL FORMED BY OUR ENVIRONMENTS, experiences, and opportunities. In my case, those influences are rooted in my childhood in sub-Saharan Africa. I'm descended from members of the rural West African Gola tribe as well as freed slaves and free Blacks who, with the aid of James Monroe, Francis Scott Key, and other members of the American Colonization Society, crossed the Atlantic aboard the USS *Harriett* in 1829 from Norfolk, Virginia, bound for the new colony of Liberia. The colonization of free Blacks in Africa was a divisive issue among Blacks and Whites in the 1800s, and prospects for the success of Liberia were uncertain. However, in the decades following its independence in 1847 from the American Colonization Society, my ancestors experienced Liberia as one of the most promising countries on the continent, fueled by income from natural resources, including rubber, iron, and timber, that bolstered its economic development. You can still see vestiges of that wealth today in the iron ore, gold, and diamond mining operations and a few beachfront resorts, but you'll also witness the extreme poverty resulting from decades of conflict and its attending economic collapse, leaving people able to earn a mere $900 average annual household income as of 2017—one-third of which came from Liberians living abroad.

Even as a child in the capital city of Monrovia in the 1960s and 1970s, I could see how opportunity and privilege—or the lack of them—shape each individual's trajectory. I was keenly aware of how fortunate my family was in terms of our quality of life and good health. From inside the safety and comfort of the car I traveled in to school and my piano and ballet lessons, I watched little children even younger than I was carrying their siblings on their backs, walking barefoot, and needing to contribute to their family's incomes by selling things on the street.

While I lived in a comfortable home that overlooked the Atlantic Ocean, many of these children lived in small, crowded homes with dirt floors; I'd see them carrying buckets of water on their heads because there was no plumbing. Most children went to schools that had few chairs and desks, much less the books, equipment, and well-trained teachers that my school had, right in the same city. It felt unfair that I'd been blessed with safety and security while those around me suffered from poverty and poor health. The disparities were all around me—they were hard to overlook.

My mother, Izetta Roberts Cooper, and my father, Henry Nehemiah Cooper, both reinforced this awareness of unfairness while teaching me and my brother and sister the importance of professionalism and service to others. My father was a surgeon; my mother was a librarian. Both worked tirelessly to improve the lives of others, especially those in disadvantaged positions. They told us that to those whom much is given, much is expected.

My parents, who were childhood friends, came to the United States separately in the late 1940s to attend college. My mother earned her degree in education at Boston University before going on to the master of library science program at Case Western University, while my father earned his degree in chemistry from Clark College in Atlanta before moving to Nashville to attend Meharry Medical College in 1950. He and my mother were married in 1953 when he was a third-year medical student. My father experienced the deeply embedded racism that all African Americans faced in the South then—and still encounter now. He, like his fellow Blacks, was thwarted by "Whites Only" signs, as when he went to shop for a new clarinet. He was once stopped by a police officer at night while driving, and he feared for his life during that Jim Crow era, when many African Americans relied on *The Green Book*, a guide that provided them with advice on safe places to eat and sleep as they traveled through America. The officer said to him, "Boy, don't you know that when you leave Atlanta you're in the state of Georgia?" making the point that outside of those city bounds, if a Black man was accused of wrongdoing by a White person, his fate was sealed, since his voice and life were not worth much in most of the Southern states, especially outside of urban areas. These and many other examples of both oblique and blatant racism occurred throughout my parents' decades in the United States, and it further deepened their empathy and instincts to help others who were disenfranchised—yet it never seemed to impact their openness to relationships with people of all races.

After earning his medical degree, my father did his residency in general surgery at Meharry, then went on to a surgical oncology fellowship at Memorial Sloan Kettering Cancer Center in New York, before returning to Meharry to serve as chief resident in surgery. In 1961, with my brother and sister in tow, my parents returned to Monrovia—much to the pleasure of the Liberian government, which was sorely in need of trained professionals. I was born a couple of years later. Not long after their homecoming, my father became the attending surgeon at the Liberian Government Hospital in Monrovia. Between 1961 and 1972, he served on the staff of various hospitals and eventually became the personal physician to the president of Liberia, William R. Tolbert Jr. While helping countless patients, my father's efforts helped drive the establishment of an unprecedented medical infrastructure in Liberia.

Meanwhile, my mother was appointed head librarian at the University of Liberia in Monrovia, a position she held until 1966. Like my father, she focused her energies and talents on systemic improvements. She introduced the Dewey Decimal System into the library, developed an interlibrary system within the country, ensured that the library's holdings supported the college curriculum based on faculty and staff recommendations, and established the African Room to house books primarily by and about Liberians as well as books on Liberia and Africa in general. She also hosted a television program called "The World of Books" and helped to set up the library for Liberia's longest-serving president, William V. S. Tubman. While she was contributing all of this and more to our country, she was simultaneously instilling a love of reading in me. And, through observing her work with data organization, collections management and cataloging, I learned about research practices, leadership, and information management, all of which have served me well throughout my career.

My parents taught my siblings and me that life was about how you use your skills and talents to improve the lives of others. In 1962, they established the Cooper Clinic for Diagnosis and Special Surgery in Sinkor, Monrovia; President Tubman delivered the dedication address. The clinic was a small hospital that provided primary care and specialty services to a broad cross section of the community. My father oversaw the day-to-day operations, while my mother assisted at the clinic by managing purchasing and the dietician staff. During his time there, and as the first chief medical officer of the John F. Kennedy Hospital in Monrovia, my father saved countless lives and delivered many babies. Through his deep connections with our indigenous relatives and their culture, and his command of many of the other local languages, he was particularly adept at bridging the divide between the more Westernized, urban, and affluent groups and the indigenous, rural, and often less economically advantaged communities in Liberian society. Witnessing the results of his work taught me the importance of listening to and respecting patients from all walks of life.

Beyond their service through their professions, my parents were also volunteer community activists. My mother was on the leprosy control board that fought to eradicate what is now known as Hansen's disease. In 1977, my father founded the Liberian Cancer Society, which was later incorporated into the International Cancer Society. I spent countless hours listening to them talk about these initiatives, as well as working at the Cooper Clinic as a pharmacy assistant, so from a young age I knew I wanted to study medicine and follow in my parents' footsteps of giving back.

Through all of their efforts, my parents illustrated the value of relationships. Family is a vital part of Liberian culture; we saw our aunts, uncles, and cousins regularly, and we came to understand the value of loyalty and respect for our elders and the contributions they made to our country. In addition to my biological siblings, my parents fostered several children from parts of Liberia where the schooling was not as strong as in the capital; these children lived with us as family. My parents emphasized the importance of maintaining and strengthening those bonds, and family remains very important to me. They also introduced us to their friends from all over the world—people whom they'd met while in school in the United States and whom they'd come to know through their professional, cultural, and civic activities in Liberia, including diplomats, missionaries, and business executives, and we socialized with children from these diverse backgrounds. Through all of these experiences that my parents brought into my life, I acquired a comfort level working with people from different cultures that would serve me well decades later in my medical and research careers and community outreach efforts.

A Violent Homecoming

For most of my childhood, we lived in Congo Town, a suburb of Monrovia, and I attended the American Cooperative School nearby. However, since the school was not yet accredited for US colleges, my parents made the decision to send me to one in Geneva, Switzerland, to aid in my being accepted to an American university, as they had both been.

The boarding school world was a different culture from the one I'd grown accustomed to in Liberia. My friends were from Germany, Iran, France, England, Turkey, and the United States. It was a multilingual, international place where people with different backgrounds intermingled—with lots of diverse perspectives to be navigated—including our tastes in music, food, dancing, and beliefs about what constituted cool or uncool behavior. I also experienced how it felt to be a minority. In Liberia, although I'd attended an international school, I was in my home country, which was led by people who looked like me from its inception. Now I was in Europe, clearly "different" and an "outsider."

During my senior year in 1980, I returned to Liberia for a cousin's wedding. Tensions were rising. Liberia had long experienced conflicts among different tribal groups as well as between the descendants of African Americans who founded the country and indigenous Africans. Until 1980, a small group of descendants of slaves and free Blacks, infamously called "Americo-Liberians" by foreigners or "Congo People" by indigenous Liberians, held power and controlled the government. But on my seventeenth birthday, April 12, 1980, the government of Liberia was overthrown in a bloody coup. That day, our American-descended president Tolbert—while still my father's patient—was assassinated by a member of the Krahn tribe named Samuel Doe, throwing the country into turmoil. Doe immediately established a military regime called the People's Redemption Council and garnered early support from a large number of indigenous ethnic groups that had been largely excluded from power since the country was founded. Doe dealt violently with the opposition, in part fueled by his fear of a countercoup. During this time, soldiers went from house to house terrorizing people, especially those who were considered to be the privileged class of Liberians, like my family. A number of my relatives and friends were raped, killed, or otherwise victimized. Many of my friends' parents were murdered because they were either involved in government or because they belonged to the middle or upper class, or were considered to be not fully Liberian because of their ancestry from other parts of Africa or America.

My father was arrested while driving to the hospital to treat victims of the violence. He was stopped, pulled out of his car, and beaten by soldiers because they saw him as one of the privileged class. I was petrified—I remember thinking that we were all going to die, that I would never grow up. I would never achieve my dream of becoming a doctor and coming back home to help make things better for those children who didn't have the good life that I'd had. And for the first time, I understood what it was like to be in a minority group without the power to protect myself. Suddenly, I was on the other side of the equation. Experiencing such powerlessness and being surrounded by such violence had a pivotal impact on my understanding of how social injustice harms everyone within a society over time, and in particular harms the relationships among groups that experience different levels of advantage and disadvantage.

A Descendant of Slavery Arrives in America

We couldn't have fathomed it then, but things would become even worse in Liberia. The 1980 coup began 23 years of intermittent unrest and eventually led to civil war that wreaked havoc on the infrastructure and impoverished the nation. Up to a quarter of a million people were killed in those years, while thousands more were forced from their homes. Fortunately, my family survived that first explosion of violence; my parents acted quickly, allowing us to leave for the United States in 1980. A letter of acceptance to Emory College, arriving at the US Embassy in Liberia, facilitated my getting a student visa and traveling to America, alone at seventeen, just weeks after the coup, with my most valuable belongings sewn into my clothes for safekeeping. The rest of my family arrived in the following months.

Although I'd vacationed in the United States with my family before, I'd never lived here. So when I arrived, I had to make some adjustments and find my place within the broader culture—in the same state where my father had been threatened by that police officer 20 years before. I found a sense of kinship with African Americans in the South because so many of the customs reminded me of my family and friends at home. And I began to understand what it meant to be Black in America, because when people saw me, they didn't differentiate me in any way from my African American friends. Rather, they simply saw a Black woman. I studied hard at Emory College for four years and received a bachelor's degree in chemistry. I truly felt those four years saved my life—not only figuratively—and for that reason I remain a loyal alumna.

After graduation, I went on to the University of North Carolina for medical school, and then completed my internship and residency in internal medicine at the University of Maryland Medical Center, a large inner-city hospital in Baltimore. Throughout my training, I expected that I'd return to Africa to make a difference within the community in which I'd been raised. But when I completed my residency in 1991, the ongoing civil war in Liberia prevented my return. Moreover, after so many years of training, Baltimore had become my new home. During my residency, I realized that the conditions in which many people were living in Baltimore paralleled what I'd witnessed on the streets of Liberia. People faced similar financial struggles, neighborhood safety issues, and reduced job opportunities. Many had to choose between paying for food, their children's educational needs, or their medication. Many had health problems that persisted from generation to generation. Just as I'd been inspired as a child to help the children struggling on the streets of Liberia, I now felt inspired to help the underserved people with whom I interacted on a daily basis in Baltimore.

One thing in particular caught my attention during my residency: I became aware of how many problems were related to differences in life experiences between patients and health care providers. Many of these social gaps were similar to those I'd seen between poor people in Liberia and my upper-middle-class family and our international friends. Disconnects in communication, trust, respect, and understanding were undermining the health care of patients from underserved communities, who were overwhelmingly African American.

These experiences led me to pursue further training in public health. I attended the Johns Hopkins School of Medicine, where I completed a general internal medicine fellowship and obtained a master's degree in public health from the Bloomberg School of Public Health. After completing my fellowship in 1994, I joined the university faculty, which I've been a part of ever since. I began my career as a practicing internal medicine physician, but soon my roles expanded to include those of a clinical and public health researcher, mentor, teacher, leader, and a social justice advocate and activist.

During my early years on the Hopkins faculty, some of my colleagues discouraged me from working in the area of health disparities. They thought that line of research lacked prestige, and that I'd be unlikely to find sources of funding to support my work. Many people at my institution and in the medical field at that time didn't see health or health care disparities as a major issue to be addressed by academic medicine, but rather more by public health and the social sciences. There were many challenges—not enough administrative support, rejections from journal editors, harsh reviews of grant proposals, and doubts about the rigor of my research and the broad relevance of the topic. With the encouragement of a precious few including my mentors Daniel Ford, David Levine, Debra Roter, and Neil Powe, I persevered, staying focused on my goals and committed to my beliefs, and eventually met with great success by the standards of academia. When I was promoted to full professor in 2007, I became the first woman of African descent to achieve that distinction at Johns Hopkins University School of Medicine—more than 100 years after its founding.

Being the first of anything creates both opportunities and challenges. The people I've met along this journey have taught me so much, and I want to share those lessons with you in this book. I'll write about what drew me into this field, about my transition from local work in Baltimore to global work back in

West Africa, and about what we can all do to work together to end health disparities for the common good. But before we begin, it's important to understand a bit about those who taught me, the context of my work, and why health equity is so important.

Receiving and Giving through Mentoring

Throughout my life and education, I've been incredibly blessed and guided by a diverse group of mentors and role models. They've come from many different racial backgrounds, age groups, genders, religions, and specialties. I've needed all their values to teach me what I know about science, medicine, relationships, and life. They also taught me the importance of leading by example through demonstrating my values and behaviors to those who will follow me. Dr. Ford taught me the importance of having a specific focus—something that would make my work stand out from that of others—and of being a resourceful and generous collaborator. Dr. Levine encouraged me to stay true to my values through my behaviors and modeled the competence, benevolence, and advocacy that would engender trust from my colleagues and community partners. Dr. Roter set the example for the enthusiasm, optimism, and resilience I would need in my career, the fearlessness in pursuing questions others had not contemplated, and the balanced approach to family and self-care. Dr. Powe modeled the importance of determination, perseverance, scientific rigor, and a skilled approach to leveraging professional networks.

You'll hear more about one of my influential role models, Elijah Saunders, in chapter 1. Another pivotal role model on my journey was Levi Watkins Jr. Dr. Watkins grew up in Alabama and, as a teenager, served as a driver for the Reverend Martin Luther King Jr. He was the first African American to be admitted to and graduate from Vanderbilt University's School of Medicine and the first African American surgical resident at Johns Hopkins. In 1980, he implanted the world's first automatic defibrillator in humans. He was a civil rights activist who dedicated his entire life to achieving social justice, equity, and inclusion in medicine in his community and in our larger society.

I first met Dr. Watkins at a welcome reception that he hosted for incoming underrepresented minority students, residents, and postdoctoral fellows. This was an institutional tradition he had begun early in his tenure at Johns Hopkins that continues to this day. A few years later, when I applied for the Robert Wood Johnson Foundation's Amos Medical Faculty Development Program (a prestigious program created to increase the number of faculty from historically disadvantaged backgrounds who achieve senior rank in academic medicine, dentistry, or nursing and who can foster the development of subsequent generations of scientists), I sought him out, knowing that he served on the National Advisory Committee. He shared his wisdom with me, supported me through the application process, and continued to advocate for me throughout my career.

Dr. Watkins inspired me to continue to pursue my professional goals in health equity research at a time when it was difficult to be funded to do that kind of work, and when this area of research was not yet recognized as a scientific discipline. He also taught me a lot about faith and courage. He was outspoken and encouraged me to call out the injustices I saw in health care. Yet he also let me know that he was familiar with fear. He'd grown up in a time when it wasn't safe to speak out about issues like racial discrimination and social injustice. He said and demonstrated through his actions that if you had faith in yourself and also in something bigger than yourself, you could move past any fears you had. When he died unexpectedly in 2015 at the age of 70, I resolved to honor his memory and carry on his legacy.

It's critically important in the world of medicine and public health to pass on information, skills, and ways of thinking to the next generation who will carry the torches forward to heal patients and whole communities in the future. Just as I looked to echo my parents' contributions to medicine and society, my mentors and role models inspired me to become a devoted mentor. In addition to providing me with a powerful professional network of peers and role models across the country, my selection as a 1995 scholar in the Robert Wood Johnson Foundation's Amos Medical Faculty Development Program provided four years of salary support to protect my time for research and additional funding that solidified my foundation as a leading health disparities researcher and mentor for future generations. The National Advisory Committee members and specifically, two persons from that program, James R. Gavin III, a renowned physician scientist in the area of diabetes and National Program Office Director from 1993 to 2013, and Ruby Puryear Hearn, a biophysicist, one of the creators of the program, and a senior vice president at the foundation, remain my staunchest professional role models and sponsors. From 2006 to 2016, a Midcareer Investigator Award in Patient-Oriented Research from the National Institutes of Health provided additional funding toward this effort, as the award does for other mid-career and senior scientists by protecting their time for research and mentoring. I also participated in an innovative program called Culture Change in Academic Medicine (C-Change) at the Leadership and Mentoring Institute at Brandeis University, which provided excellent training through a year-long group peer mentoring approach and time to envision writing this book. All of these experiences prepared and motivated me to mentor more than 75 individuals from various disciplines and clinical specialties and at various levels of training over the course of my career. Through mentoring, I not only pass along what I've been taught and what I've discovered, but I also, through the knowledge and experiences of my mentees, stay current and develop a broader understanding of the field through their varied interests, skills, and connections.

How Health Disparities Affect Everyone

I began writing this book in the first half of 2020 when the COVID-19 pandemic spread across the world, bringing to light health disparities that have existed among minority and at-risk communities worldwide for many generations. Another trauma burst into view in the United States a few months later sparked by the brutal killing of George Floyd in Minneapolis, which set off worldwide protests against systematic racism. Like other health disparities, the epidemic of Black men and women dying violently at the hands of police is one of very long standing. The extent of anti-Black racism becomes evident when people see an unarmed African American man suffocated by police, but many other inequities are less obvious. By the fall of 2020, more than a million lives had been lost around the world due to the pandemic—more than 200,000 in the United States alone—and racial tensions persist.

Both crises fostered a sense of community bereavement—a collective, cumulative grief—here and abroad that in turn prompted civic actions like marches and deeper examinations of everything from police apprehension tactics to the designs of state flags to "vaccine nationalism"—when countries, particularly those with more wealth, secure doses of vaccines for their own citizens before the vaccines are made available in other, typically poorer, countries. They demonstrate the essential need for societies to achieve not just health equity but social justice as well. Before the pandemic, minority communities were already disproportionately affected by systemic racism. They endured a confluence of factors—not having access to healthy food or effective education, multiple generations living crowded in small houses, having to rely solely on public transportation to travel to essential jobs that pay low wages and offer few benefits, and having less access to health care, both financially and geographically. This history of institutionalized, sanctioned mistreatment made minority populations especially vulnerable when the COVID-19 pandemic struck.

Yet the pandemic and the protests also revealed how interconnected and vulnerable we *all* are. Because we've all felt their impacts on our daily lives and freedoms, as well as on the economic stability of the entire country, they showed that our well-being depends as much on those around us as on ourselves. The striking racial and ethnic disparities reported for COVID-19 (discussed in more depth in chapter 7) are a clear reminder that failure to protect the most vulnerable members of society not only harms them but places everyone at risk. COVID-19 disparities are not solely the responsibility of those experiencing them. People from high-risk groups live in neighborhoods where opportunities to access health care and obtain safe housing and healthy food are limited. Essential workers, many of whom are from these groups, often don't have benefits such as sick leave and health insurance, which would permit them to self-quarantine or to access health care without penalties. Disparities reflect social policies and systems that we, as a society, have allowed to create these everyday injustices that become magnified in times of crisis.

The United States and other countries must develop a new kind of "herd immunity," where all racial, ethnic, and social class groups become protected from the effects—those easily witnessed, and those less visible—of societal discrimination and disadvantage.[1] Physical and emotional well-being are human rights—everyone deserves the opportunity to live a healthy life. The basic needs we all require for this are stable housing, safe environments, food security, access to a good education, jobs that pay a living wage, and high-quality health care. My hope is that through reading this book, your awareness will be sharpened so that you'll be able to see the injustices around you and call them out. It may not be possible for one person to reverse the tide, but together, through conversations that prompt policy-shaping actions, we can raise our collective consciousness. The great seal of the United States includes the motto *e pluribus unum*—out of many, one. We're each a part of America, and we all deserve equal treatment across all available opportunities and hard-won rights. But each of us must also contribute to making that vision a reality.

Why Are Health Disparities Everyone's Problem?

INTRODUCTION

Five Miles Apart

IMAGINE TWO 60-YEAR-OLD WOMEN, one African American and the other White. The first woman, Anita Jones, lives in Baltimore's Latrobe Homes neighborhood. Anita's husband, who had worked as a mass transit bus driver, died of a heart attack several years ago, and ever since, she's struggled to make ends meet. She graduated from high school but never went to college, so it's been hard for her to find a good job that also pays benefits. Instead, she has two part-time jobs—one as a food service worker at a downtown hospital, the other as a call center telemarketer—that together pay her $35,000 a year. While that salary never stretches far, she now supports her daughter and grandson, who recently moved into her small apartment after her daughter lost her job.

Anita rarely goes to doctors. Neither of her jobs provides health insurance, and she's both too young to receive Medicare and slightly over the income threshold for Medicaid. And with two more mouths to feed, she had to give up her Affordable Care Act insurance, which cost hundreds a month. It's hard to find the time to go anyway, given juggling two job schedules. A doctor at her hospital's emergency room put Anita on medication for her high blood pressure after she was feeling faint during a shift, but she takes it sporadically. She's not entirely convinced that she needs it—she feels better—and she tries to save the money because without insurance, the medication costs $20 a month. That same doctor also told her that she's prediabetic and needs to manage her diet if she doesn't want to get diabetes. She knows she should eat healthier, but cooking old family recipes on her rare days off is one of her biggest pleasures outside of church. Making those dishes is a way to show her love to her daughter and grandchild, even if her jobs usually prevent her from sitting down to dinner with them.

A WORLD AWAY

Deborah Clark, a White woman, lives less than five miles away from Anita in Baltimore's affluent Roland Park neighborhood. Her husband works for J. P. Morgan, where he earns $225,000 a year. Deborah was a stay-at-home mom when they were raising their sons, but now that they're grown she spends her free time volunteering at the library a few blocks away and organizing fundraising events for the hospital where Anita works. In fact, although they've never met, Anita has served food at some of Deborah's parties to earn a little extra money.

Deborah receives her health insurance through her husband's job and regularly visits doctors. Like Anita, she's overweight and suffers from high blood pressure. She takes two medications to control her blood pressure, and although one of them has some side effects, they're nothing she can't handle. Five years ago, she was diagnosed with breast cancer after her annual mammogram revealed a lump. Fortunately, they caught it early, and she received surgery and radiation at one of the best cancer centers in the country. She's been in remission ever since and has made her health a priority. She goes on a three-mile walk through the park every day, buys lots of fresh fruits and vegetables, and rarely goes out to eat. Both she and her husband belong to the same gym—he's overweight, too—although neither of them uses it often. But when Deborah goes in for her twice-yearly checkup, she's told that she's in excellent health. She's more worried about her husband's weight because his father died of a heart attack, but she feels better when the doctor—an old friend of the family—takes the time to discuss a new clinical study for a promising weight-loss program that her husband might qualify for.

Back to East Baltimore: Since that ER visit, Anita has been trying to lose some weight and get a little exercise, but she hasn't been successful. The only place to buy groceries in her neighborhood is a corner market along her walk to and from the bus stop that stocks mainly canned and prepackaged goods. She likes to walk, but she doesn't feel safe in her neighborhood (police sirens are a regular part of life) and there aren't any parks nearby. The meals that are a perk of her job at the hospital are convenient and she's glad to save that money, but the food tends to be loaded with excess salt, sugars, and fats.

Anita's health grows steadily worse over her sixtieth year, and finally her daughter convinces her to see a doctor at a clinic that offers a sliding-scale payment option. After a long time in the waiting room, Anita is seen by a young White doctor. Although he's polite and asks her a few questions at the beginning of the appointment, he spends most of the time insisting that she needs to take her blood pressure medicine. He also tells her that at the rate she's going with her weight, "you'll have diabetes by the end of the year." Anita tries to talk to him about why she can't take her medicine, but leaves the visit feeling worse about her health and her circumstances than ever before and guilty for not being able to stick to a better diet. The doctor seems a little distracted and in a hurry to end the consultation, so she forgets to tell him she's been getting headaches a lot more than usual, too.

A month later, Anita falls to the floor as she gets out of bed. The right side of her body is weak, and she has a numbness and tingling on that side as well. Her daughter is taking her grandson to school, so she's all alone. Although she considers calling for an ambulance, she knows that they could never recover from the cost of being treated in the emergency room again. She manages to get to the kitchen, where she takes two of her blood pressure pills and then sits in the armchair by the door, praying until her daughter gets home and can help her figure out what to do. But by the time her daughter arrives, Anita is unconscious. Her daughter calls 911, but the paramedics arrive too late, and Anita is pronounced dead at the scene. Cause of death: stroke. She was 61 years old.

A DIFFERENCE IN OUTCOMES

One afternoon at their grandson's baseball game, Deborah's husband clutches his chest and his face turns beet red. "Call 911," he says. The EMTs arrive within a few minutes and are able to restart her husband's heart with the defibrillator on the way to the hospital.

As Deborah often remarks later, it was the wake-up call that her husband needed to pay attention to his health once and for all. He has a procedure to remove the blockage in a major blood vessel in his heart and spends a week recovering in the hospital, all of which is paid for by his insurance. When he gets home, he slowly but surely loses the weight, helped by a physical therapist, a nutritionist, and his general practitioner, all of whom he sees by paying $25 copays. He lives until he's 83. Deborah passes away peacefully at 91 with her sons and grandchildren at her side.

Although they lived just five miles from one another, Anita and Deborah's stories illustrate the breadth and depth of the health disparities that exist in the United States. The two women and their husbands suffered from similar health conditions, but their outcomes were vastly different due to the significant discrepancies in their life circumstances, the types of health services that were available to them, the care they received, and the manner in which they received that care.

I've worked in Baltimore for 30 years now. For the past several years, I've seen patients at the East Baltimore Medical Center in Madison East End, where Anita worked and lived. Home mostly to racial and ethnic minority groups, the neighborhood has an average household income of $30,389, a life expectancy at birth of 64.8 years, and a homicide rate of 46.3 per 10,000 people. Just five miles away in Roland Park, where Deborah lived in a predominantly White, wealthy Baltimore neighborhood, the average household income is $90,492, the life expectancy at birth is 83.1 years, and the homicide rate is 4.1 per 10,000 people. These differences are hard to comprehend—and completely unacceptable. How can their life expectancies differ by 20 years when Anita and Deborah live so close to one another? Yet the same is true for many cities, not just in the United States, but around the world.

HEALTH DISPARITIES AND HEALTH EQUITY

Health disparities like these have become a focus of academic, medical, and political attention over the past several decades.[1] In the United States, the 1985 Heckler Report on Black and

Minority Health marked the first time that the federal government convened a group to conduct a comprehensive study of racial and ethnic minority health.[2] Under the leadership of former Secretary of Health and Human Services Margaret Heckler, this landmark report elevated minority health to the national stage by pointing out that Blacks and other minorities disproportionately experience more deaths from certain diseases than do Whites, in part because of inequities in their health care. The report presented many recommendations, including:

- emphasizing prevention and primary care;
- expanding prenatal Medicaid benefits;
- increasing minority communities' access to health professionals and medical care;
- increasing collaborative efforts of state and local governments, professional associations, and other health-related organizations;
- improving data quality; and
- a research agenda that would increase risk factor identification, focus on health education, and elucidate the relationship between sociocultural factors and health.

The Heckler Report served as a catalyst for the creation of the Department of Health and Human Services' Office of Minority Health (OMH) in 1986 and the Centers for Disease Control and Prevention's office focused on minority health in 1988 to drive programs and policies that eliminate or reduce health disparities. The report also led to the funding of research studies such as the Strong Heart Study, a large study of cardiovascular disease and its risk factors among native Americans, and the Jackson Heart Study, the largest study in history to investigate the risk factors for high blood pressure, heart disease, stroke, and diabetes in African Americans. Later, the Food and Drug Administration and many states and territories would establish offices focused on minority health.

In 1993, the Health Revitalization Act established the Office of Research on Minority Health which built upon the work of the OMH to support a better organized, more collaborative, and more financially aware approach to studies of minority health. It also worked toward the inclusion of members of minority groups as subjects in such clinical research. At that time, most of the science focused on documenting the existence of disparities in health and health care by race. When the term *health disparity* came into use in the United States in the early 1990s, it was intended to indicate a specific kind of difference in health between groups—that is, worse health among socially disadvantaged groups, such as racial and ethnic minorities and socioeconomically disadvantaged persons of any racial or ethnic group. For many years, however, the term was used more generally to describe differences in health between population groups without further specification.

In 2000, under the leadership of President Bill Clinton, the Minority Health and Health Disparities Research and Education Act was passed, which established the Center for Minority Health and Health Disparities at the National Institutes of Health and the Healthy People 2010 agenda. Under the leadership of US Surgeon General David Satcher, the two overarching goals of Healthy People 2010 were to increase quality and years of healthy life and to eliminate health disparities. By 2010, when Healthy People 2020 was released, the term *health disparities* was used more specifically to denote a health difference that is closely linked with economic, social, or environmental disadvantage. Between 2010 and 2020, many strides were made in the health of Americans: life expectancy at birth increased, and rates of death from coronary heart disease and stroke decreased. However, it's apparent that many public health challenges remain and that significant health disparities persist.

Meanwhile, in the United Kingdom, the Black Report (named for Sir Douglas Black, president of the Royal College of Physicians) in 1980 was the first official government report in the developed world to document the extent of health inequalities between rich and poor.[3] This report found that the more income people earn and the higher they are on the social scale, the better their health. Margaret Whitehead, a professor of public health, wrote a follow-up to this report. She defined health inequities in 1995 as health differences that are avoidable, unnecessary, and unjust.[4] Health inequities is the term used in most countries, where it's generally assumed to refer to socioeconomic differences in health. In 2005, the World Health Organization (WHO) established the Commission on Social Determinants of Health to draw attention to the conditions that underlie inequitable health differences. These determinants are the "conditions in which people are born, grow, live, work and age," the commission wrote. "These circumstances are shaped by the distribution of money, power and resources at global, national and local levels."[5] The commission's final report was completed in 2008, stating "within countries, the differences in life chances are dramatic and are seen in all countries—even the richest." This wasn't the first examination of the role of social and environmental factors in health, but in the past, global health agendas tended to alternate between a focus on biological and technological approaches to medical care and public health developed in siloes, and on the other hand, health solutions developed through broad multisector societal collaborations. The commission's purpose was to reinforce the latter, broader understanding of health and revitalize the WHO's constitutional commitments to health equity and social justice.

The WHO defines health equity as "the absence of avoidable or remediable differences among groups of people, whether those groups are defined socially, economically, demographically, or geographically or by other means of stratification." Put simply, health equity is when everyone has fair access and opportunities to achieve optimal health and no one is denied those opportunities because of socially determined circumstances.

The WHO also states that health *inequities* "involve more than inequality with respect to health determinants, access to resources needed to improve and maintain health or health outcomes. They also entail a failure to avoid or overcome inequalities that infringe on fairness and human rights norms."[6] Health inequities are the result of disparities, which are avoidable inequalities between groups of people within countries and between countries. Health disparities negatively affect groups of people who have systematically experienced greater social or economic obstacles to achieving good health based on their racial or ethnic group, religion, socioeconomic status, gender, mental health, cognitive, sensory, or physical disability, sexual orientation, geographic location, or other characteristics historically linked to discrimination or exclusion.

THE SOURCES OF HEALTH INEQUITIES

Health equity seems simple enough, right? It only makes sense that everyone should receive a fair opportunity to remain or become well. However straightforward it seems, the fact is that health inequities exist in nations around the world, although they take many different forms. They result from complex, interrelated, and long-standing disparities, and their elimination is similarly complicated. These disparities include socioeconomic factors, environmental factors, discrimination, cultural factors, health-risk behavior, biological factors, access to health care, and quality of health care.

Although health equity may sound the same as health equality, in fact these terms imply different outcomes. *Health equality* means that everyone receives the same treatment—for example, providing all patients who have hypertension with pill boxes to help them remember to take their medication and home blood pressure monitors that transmit readings to the physician through a mobile app. However, individuals with lower levels of education, income, and familiarity with technology might need more assistance to use these tools effectively, including perhaps a community health worker to teach them how to measure their blood pressure correctly and use the app and financial vouchers to cover the costs of broadband access or cellular service. *Equality* aims to promote fairness, but it can only work if everyone starts from the same place and needs the same things. *Equity* means that everyone is given what they need to be successful, even if those opportunities come in different forms and are used in different ways. To achieve health equity, underserved communities may require more or different resources than more advantaged communities, not just equal resources, to have the same opportunities for good health that the rest of the population enjoys.

It's commonly believed that if everyone has equal health care resources, each one of us will be healthy, but equality isn't the same thing as equity. This graphic from the Robert Wood Johnson Foundation's Culture of Health Initiative shows that "when it comes to expanding opportunities for health, thinking the same approach will work universally is like expecting everyone to be able to ride the same bike." We won't achieve equity by giving everyone the same thing or starting them from exactly the same place, but by giving each person or group the resources they specifically need for optimal health and well-being.

Source: Robert Wood Johnson Foundation

The distinction between health equality and health equity may seem abstract, but the difference is critical in public health, with life-altering and life-threatening implications for disadvantaged communities. Striving and planning for equity can help ensure that resources are directed appropriately and support the continuous process of meeting people where they are. Health equity is not just an important moral goal. Policies that promote equity can help, directly and indirectly, reduce poverty, foster social cohesion, and reduce political conflict. Higher rates of chronic and costly conditions, combined with high rates of uninsured individuals among lower socioeconomic and minority groups, result in a greater reliance on emergency services, higher treatment costs for which communities bear the burden, and a financial strain on providers and tax-funded government programs. Health equity, paired with preventive medicine and early interventions, can save both money and lives. Take, for example, curb cuts for disabled persons using wheelchairs—these also help parents using strollers, bikers, and people using grocery carts. So a solution aimed at one group facing inequity actually may end up helping many others.

Furthermore, as I'll emphasize throughout this book, just as solutions for one group can benefit many others, health disparities are everyone's problem—we're all connected, as the COVID-19 pandemic has so clearly illustrated. Systemic injustices (and everyday injustices or microaggressions) prevent everyone from having a fair opportunity to live the healthiest life possible, both physically and psychologically. These issues intersect with many other societal problems, including structural racism, socioeconomic disparities, lack of affordable and safe housing and healthy food, negative environmental exposures, and discriminatory institutional practices. If anyone is denied the best opportunities to be healthy, people across all groups, not only the immediately impacted group, can be harmed. I'll delve more into this issue, including my personal experiences with how this manifests, in later chapters. For all these reasons, achieving health equity will require a universal commitment to social justice.

In 2006, US Surgeon General David Satcher wrote, "We can achieve health equity in America, but first, we all must care enough, know enough, do enough, and persist long enough."[7] This statement has inspired me to stay the course even when the journey has become difficult. Caring, knowing, and persisting, though, are not abstractions; these active commitments become critical in our daily recognition of the impacts of these disparities on individuals—and our role in addressing them.

How Is Racism a Public Health Issue?

A large body of scientific evidence links various manifestations of racism to poor health among the members of communities impacted by it. Experts from public health and the social sciences describe three major pathways that link racism to inequities in society and health: cultural racism, institutional racism, and individual racism.

Cultural racism is the weaving of ideas about the inferiority of Blacks and other people of color into the beliefs, values, and social patterns of our society. Cultural racism leads to:

- negative stereotypes and prejudices;
- lack of support for policies that support equity and social justice;
- negative emotional responses among people in marginalized groups; and
- limited access to opportunities.

Institutional racism is reflected in policies and organizational structures that allow the dominant group (in the United States, White people) to withhold advantageous opportunities and resources from Blacks and other marginalized racial and ethnic groups. It manifests in poor health by reducing access to key resources, including health care, healthful food, good education, well-paying jobs, and safe environments. Examples of institutional racism include:

- residential segregation and redlining;
- lack of diversity among business, health care, political, and other sector leaders; and
- income and wealth disparities.

Individual-level racism is expressed through peoples' interpersonal relationships. It's reflected in bias and stereotyping behaviors and discriminatory treatment in several arenas, including:

- home and car purchasing;
- mortgage and other bank loans;
- employment;
- health care;
- policing;
- criminal sentences; and
- school-based disciplinary actions.

All three levels of racism are believed to contribute to poor health, although the most direct links are evident in individual-level racism, which produces psychological and behavioral responses and biological reactions within targeted persons, including:

- internalized racism or self-stereotyping;
- poorer mental health;
- higher rates of smoking and alcohol use;
- lower rates of adherence to prescribed medications and medical appointments;
- higher levels of stress hormones, blood pressure, inflammation, and weight; and
- higher rates of hypertension and heart disease, premature aging, and death.

FURTHER READING

David R. Williams, Jourdyn A. Lawrence, Bridgette A. Davis, and Cecilia Vu. 2019. Understanding how discrimination can affect health. *Health Services Research*. 54 2(Suppl. 2): 1374–1388.

Michael M. Gale, Alex L. Pieterse, Debbiesiu L. Lee , Kiet Huynh, Shantel Powell, and Katherine Kirkinis. 2020. A meta-analysis of the relationship between internalized racial oppression and health-related outcomes. *Counseling Psychologist* 48(4): 498–525.

Brian W. Simpson and Keshia Pollack Porter. 2020. How structural racism harms Black Americans' health. *Hopkins Bloomberg Public Health*, https://magazine.jhsph.edu/2020/how-structural-racism-harms-black-americans-health.

CHAPTER 1

The Personhood of Patients

LEARNING BY EXAMPLE

MY FATHER WAS MY FIRST ROLE MODEL for how a patient should be treated. Many of my friends and their parents told me how much they admired him and how he had provided just the right medication, performed a life-saving surgical procedure on them, or simply listened to them and comforted them during a difficult time. He took the time to know their families and hear their stories. While in this era of ever-more patients, ever-shorter office visits, and ever-greater documentation demands on health care workers it's harder to reach that level of knowledge, he lived the words of Sir William Osler (the famous clinician and diagnostician—and the first physician-in-chief of the Johns Hopkins Hospital): "It is much more important to know what sort of patient has a disease than what sort of disease a patient has."

Other than my father, perhaps the person who taught me the most about patient-physician relationships was Elijah Saunders, whom I met when I was a medical resident at the University of Maryland Hospital. Dr. Saunders was extraordinarily learned and had a wealth of clinical experience, yet he was always humble and respectful in his interactions with patients and colleagues alike. He spent many hours late into the evenings and on weekends pursuing the appropriate tests and consultations from specialists to ensure that his patients were getting the best clinical management. He had a life-long commitment to learning, often citing the most recent journal articles on topics relevant to a patient's care and encouraging his residents to seek the most current evidence to guide decisions.

Over the time that we worked together, I saw Dr. Saunders use many relationship-building approaches. He listened to patients and their family members with respect, solicited their concerns and opinions about diagnoses and treatments, answered their questions, empathized with them, and demonstrated his commitment to them and their well-being. Dr. Saunders taught me the importance of putting the patient first. And because he knew so much about his patients, their family members, and the communities in which they lived, he also taught me the importance of understanding patients' unique needs and circumstances.

UNDERSTANDING WHAT PEOPLE NEED

A few months after I completed my fellowship training and joined the faculty at Johns Hopkins on the clinical investigator track, I was excited to see that the health care field was beginning to place a new focus on addressing patients' needs and concerns through a movement toward "patient-centered care." The phrase had been introduced by psychoanalyst Michael Balint in 1964 to express the belief that each patient "has to be understood as a unique human-being."[1] Over time, it had evolved from a guide for individual clinicians interacting with individual patients to a comprehensive way of delivering health services for any organization; it encompassed domains such as understanding the whole person from biopsychosocial perspectives, seeking common ground or sharing power and responsibility, and the patient-doctor relationship (a therapeutic alliance with agreement on goals of treatment and tasks and a personal bond based on reciprocal positive regard).[2] I received my first grant from the Picker-Commonwealth Scholars Program, developed to contribute new talent in research on patient-centered care,[3] which launched my research career. I would still see patients and teach medical and public health students, but now it was clear that most of my time would be spent doing clinical research.

To help the socially disadvantaged groups surrounding me, I needed to learn more about what they wanted, needed, and received. How do attitudes and beliefs affect relationships, the quality of care, and health outcomes? I needed to examine the ways in which patients from African-American and other vulnerable communities viewed health and used the health care system.

To look at these questions, I began to study patients' attitudes and preferences regarding mental health. I focused on mental health because I noticed that it was one of those areas in which African American patients and their doctors seemed to have challenges getting on the same page. Patients would complain about physical symptoms such as fatigue, headaches, or pain, and doctors would end up ordering a lot of tests before realizing that the patient might have depression or anxiety. In fact, when asked directly about depression or anxiety, a lot of my African American patients would tell me, "I'm too blessed to be stressed."

My first study involved three focus group discussions related to patient experiences and concerns regarding treatment for depression.[4] The first focus group comprised seven health professionals—four physicians and three social workers—involved in the care of patients with depression; the second comprised eight White patients with a recent episode of depression; and the third comprised eight Black patients with a recent episode of depression. The study divided Black and White patients into groups to better understand how patient attitudes, preferences, and help-seeking behaviors might vary across racial groups. Questions asked during these focus groups addressed experience with depression, seeking help from health professionals or others for one's problems, treatment preferences, and perceived barriers to mental health care. Discussions were audiotaped, transcribed, reviewed, and then grouped into categories with specific themes.

Black patients cited spirituality as a way of coping with their depression more often than White patients. Black patients also discussed using their church and church members for support more frequently than White patients. In addition, the Black participants perceived stigma as a particularly important barrier to getting treatment. Many of them felt that the idea of seeking professional help for mental health problems was not culturally acceptable to them or to their family members or peers. Admitting to mental health struggles, they felt, signaled a weakness of character or inability to cope with problems that were just a part of living as a Black person in America. In the Black patient focus group, participants also raised issues of cultural mistrust and concerns about being used as guinea pigs for medical experimentation. Furthermore, Black patients cited the dearth of mental health professionals from their own gender, race, and religious background as concerns.[5] Patients of both races saw health professionals' technical skills and interpersonal skills as important in their decisions of whether to disclose their innermost feelings and to accept any recommended treatments.

I then did studies with larger numbers of patients from different parts of the country.[6] I found that, compared with Whites, African Americans, Asians/Pacific Islanders, and Latinos were more likely to prefer counseling to medications, less likely to believe that medications were effective and that depression was biologically based, and more likely to believe that antidepressants were addictive and that counseling and prayer were effective in treating depression. One of the most intriguing findings was that patients, regardless of race or ethnicity, told us that their relationships with doctors and other health professionals were the most important reasons they ended up accepting treatments and adhering to recommended tests and medications. Along with the specific attributes of medications and counseling, the most important aspects of care were being able to trust the health professional to act in one's best interests, having information and knowing what to expect from treatment, and health professionals' interpersonal skills (knows their patients, listens, understands patients' problems, approaches patients as individuals, makes patients feel comfortable, supports and encourages patients), and physicians' recognition and validation of patients' depression (recognizes depression, believes patients' symptoms are real). Slightly lower on the list of concerns were the affordability of mental health treatment and having health insurance to cover it. These early studies gave me a better understanding of how cultural and social factors might influence willingness to both seek and use health care and other health behaviors of patients from communities of color. They also piqued my interest in better understanding how patient-physician communication and other aspects of the patient-physician relationship, such as trust, might impact health outcomes, and whether any differences in these relationships could explain the racial and ethnic disparities in health care that were starting to be revealed through the research.

THE PATH FROM MISTRUST TO TRUST

Good relationships are essential to building trust, and trust is vital to successful health outcomes. Scholars define trustworthiness as the ability to be relied upon by others based on benevolence, integrity, or the competence of persons and institutions.[7] Patients need to trust their physicians enough to share their personal histories and to follow their treatment plans. Physicians need to trust their patients enough to believe what they say and arrive at the proper diagnosis and path for recovery without making hasty or moral judgments—without blaming the patient before they fully uncover the causes of their issues, or making erroneous assumptions. Unfortunately, many members of disadvantaged groups distrust the health care system, driven by historical oppression, structural racism, disparities in care, and personal experiences of discrimination.

Trust is based on shared understanding. Disadvantaged communities witness and experience the far-reaching and devastating effects of disproportionately poor health outcomes while struggling to understand how to use the health care system to their maximum advantage. On the other hand, physicians diagnose and treat these poor health outcomes but don't necessarily understand how to tailor their treatments to reflect the underlying drivers of health disparities. As tax-paying, contributing members of society, many disadvantaged individuals believe that the health care system should address the many causes of their poor health outcomes. In contrast, though many health system leaders recognize the complex causes of health disparities, they view their role in eliminating these disparities as confined to the equal—but not necessarily equitable—provision of health care across populations they serve.

Trust in the health care system is also low among minority groups because historically underserved populations continue to have limited access to primary care. Instead, they often have to rely on more expensive and fragmented health care services, such as emergency departments.[8] In one of the early studies my colleagues and I undertook to examine this issue, we administered a telephone survey to 118 adults in Maryland and asked respondents to rate their level of trust in physicians, health insurance plans, and hospitals.[9] We found that Blacks were 37 percent less likely than Whites to trust their physicians. Although the racial difference in trust of hospitals was not statistically significant, Blacks were more likely than Whites to be concerned about personal privacy and the potential for harmful experimentation in hospitals. We concluded that these differences in trust might reflect divergent life experiences of Blacks and Whites in American society, and we called for improved understanding of these factors in efforts to enhance access to care and quality of care among African Americans.

Mistrust of the medical system has many sources.[10] The Tuskegee study of untreated syphilis is infamous; during this study, the US Public Health Service withheld treatment from 600

Black sharecroppers and their families in Alabama for nearly three decades after penicillin became the accepted treatment for syphilis. In Baltimore, many cite the experience of Henrietta Lacks, the African American woman who was the unwitting source of the HeLa cancer cell line—a defined population of cells that can be maintained in culture for an extended period for research purposes. In 1951, Lacks had a tumor biopsied during treatment for cervical cancer at Johns Hopkins Hospital. As was standard practice at the time, her physicians didn't seek her consent before using her tissue for research. Extracted cells from her tissue were cultured to create the HeLa cell line—the first "immortalized" human cell line and one of the most important in medical research; it's still in use today. At the time, there were no federal regulations or restrictions on the use of patients' cells in research; however, the cells have been used in developing the polio vaccine, gene mapping, in vitro fertilization, cancer therapies, HIV research, and more, by scientists all over the world. Mrs. Lacks's story has been known in the research community for a long time, but it became more widely known after the publication of a best-selling book titled *The Immortal Life of Henrietta Lacks* in 2010. Despite the many research breakthroughs that Mrs. Lacks unknowingly enabled, it wasn't until many years after her death that her family learned about her contributions to medical progress. In 2013, the National Institutes of Health announced that it had reached an understanding with the family of the late Henrietta Lacks to allow biomedical researchers controlled access to the whole genome data of cells derived from her tumor.[11] The policy gives the Lacks family the ability to have a role in work being done with the HeLa genome sequences and track any resulting discoveries. All researchers who use or generate full genomic data from HeLa cells are now asked to include in their publications an acknowledgement and expression of gratitude to the Lacks family for her contributions.

Within underserved communities, a lack of transparency around addressing the root causes of health disparities within the American health care system raises questions about the extent to which the system is truly committed to advancing the health of these populations. The resulting tension intensifies the historical lack of trust between health systems and underserved communities. People from at-risk groups and historically disadvantaged communities can feel that they're not being respected and are being stereotyped by clinicians and researchers, or worse, taken advantage of.[12] Building and sustaining trust between patients and their doctors, and health care systems and their communities, will be essential to improve the health of disadvantaged groups and to eliminate health disparities.

DELVING BELOW THE SURFACE

The studies that I undertook in the first decade of my career pointed to the importance of the patient-physician relationship in health disparities. Much was already known about the importance of these relationships in producing better patient outcomes, including more engagement in care, improved self-rated health, and greater adherence to treatment. But much less was known about relationships across social differences such as race, ethnicity, language, and culture.

Researchers who study the impact of culture on health often use the metaphor of an iceberg. The point they make is that the portion of a person, an organization, or a group that is visible above water is, in reality, only a small piece of a much larger whole—the tip of the iceberg. Many other facets of that person, organization, or group, while less visible, are just as essential to our understanding of how they think and behave. Visible characteristics include age, gender, social class, ethnicity, race, language, and physical fitness, among others. Invisible characteristics include beliefs, behaviors, attitudes, values, preferences, and role orientations that influence relationships. The iceberg also symbolizes risk. Improving relationships, especially across cultural and social differences, is a risky business. Issues are deep, complex, longstanding, and emotionally charged.

At this stage in my career, I knew that if I wanted to truly understand how relationships might be contributing to health disparities, I needed to delve below the surface.[13] I needed to better understand the various dimensions of the patient-physician relationship and the potential impact of several characteristics of each participant—visible and invisible—on the relationship and its outcomes.

CHAPTER 2

The Patient-Physician Relationship

DURING MY FIRST YEAR OF RESIDENCY at the University of Maryland Medical Center, I was called to the emergency room to admit a young African American woman who was about my age. I didn't know at that moment that this would be my wake-up call to delve deeper, below surface appearances and facts, to better understand the patient and the problem that needed to be addressed. All they told me was that she was in her early 20s, was HIV positive, and had come in with a fever and a headache. All kinds of scenarios went through my head. Was she a drug addict who had shared dirty needles? Had she been infected because she was sexually promiscuous?

As often happens during grueling residency training times, I was tired and stressed. It was my last night as an intern, and I was ready for that difficult year to be over. I didn't feel up to talking much with yet another patient at that hour of the night. At this point in the year, I was so used to admitting patients in the middle of the night that I could do the full work-up without thinking too much about it—or about them—beyond what medical issues they presented. I thought, let me do this quickly so I can get back to bed. But when I saw her I was immediately struck by how small, frail, and afraid she looked. She could have been one of my friends or cousins from Liberia. I didn't want to think or feel anything because I was so exhausted, but I couldn't dodge the emotional connection. She must have felt the same sense of familiarity with me, because as we began to talk, she told me more and more about her life. She had contracted HIV from her boyfriend, who she didn't know was using drugs. I was the first doctor, she said, with whom she'd felt comfortable enough to tell her whole story.

That experience was a turning point for me. I had known for a long time—as far back as my childhood, witnessing my father's example—that I needed to bring my whole self into my encounters with all of my patients, but in that moment, I realized that—whether it was due to the stress and technical problem-solving focus of residency, a lifetime of cultural and social conditioning, or all of these factors combined—I was guilty of making the same erroneous assumptions about patients of color as many of my White colleagues. And this time, I realized, it wasn't going to be enough for me just to be a good doctor to each patient, one person at a time. I had an obligation to address the health problems faced by people of color and underserved groups on a larger level. Both those realizations have shaped my interests and research ever since.

THE PATIENT-CENTERED MEDICAL INTERVIEW

The crux of medical care is the relationship between physician and patient. A physician needs to learn a patient's history in order to diagnose that patient properly. To establish an effective treatment plan, physicians have to understand that patient's individual needs and what treatments he or she will adhere to and find acceptable. If a doctor doesn't understand what's going on with a patient, the result could be a misdiagnosis or inappropriate treatment plan. And, reciprocally, patients need to come prepared with as much information as possible about what they're experiencing, what medications they're on, and their family histories. In medicine, we often discuss changes to medications, therapies, or procedures to improve health outcomes. All these possibilities have to be built on a foundation of a strong relationship. This is what we mean when we say patient-centered care.

The field of patient-centered care began as an explanation of how physicians ought to interact with individual patients. Using this orientation, the patient-centered medical interview "approaches the patient as a unique human being with his own story to tell, promotes trust and confidence, clarifies and characterizes the patient's symptoms and concerns, generates and tests many hypotheses that may include biological and psychosocial dimensions of illness, and creates the basis for an ongoing relationship."[1] I wanted to study whether differences in patient-centered communication could explain some of the racial and ethnic disparities we were observing in health care and outcomes.

These disparities exist across medical practices, but the early research began with primary care physicians who, as the principal points of medical contact, diagnose and manage most of the conditions connected to health disparity documentation. The foundation of primary care is a sustained relationship between patients and the clinicians who care for them. To meet the challenge of health disparities, primary care clinicians must provide exemplary care to individual patients in the context of their personal circumstances. It's a contract between the two, in which the patient shares and discloses as much as possible, and the health care provider listens and consults with them to decide upon a course of care.

My interest in the effects of the patient-physician relationship on health outcomes led a group of colleagues and me to conduct a study of race, gender, and patient-physician relationships that was published in the *Journal of the American Medical Association* (*JAMA*) in 1999.[2] Though many previous studies had documented gender differences in health care received by patients, and some studies had examined racial differences, few studies had examined interpersonal aspects of care, and even fewer had related those differences to the race of the physicians who were seeing the patients. The study involved telephone surveys conducted between November 1996 and June 1998 of

1,816 adults ages 18 to 65 who had recently received primary care from a large managed care organization that primarily served the Washington, DC, metropolitan area. Patients from a total of 32 practices, representing general internal medicine and family practice, were interviewed. Females made up 66 percent, Whites 43 percent, and African Americans 45 percent of the population. The physician sample was 63 percent male, with 56 percent White doctors and 25 percent African American doctors.

We focused on a measure of patient-centeredness called participatory decision-making—the tendency of a physician to involve a patient in decision-making about their care. The study found that racial and ethnic minorities rated their visits with physicians as less participatory than did Whites. Within all groups, patient satisfaction was highly associated with physicians' decision-making styles. Physicians who involved their patients more in the process of diagnosis and treatment received higher ratings from patients. These were important findings because previous research showed that individuals who deemed their medical visits less participatory were more likely to disenroll from the practice within a year.

Many studies have shown that minorities and people who are poor often receive less optimal care than do White Americans with higher household incomes. Our 1999 *JAMA* study was among the first to unearth what might be driving these differences in care. To receive proper care, health care professionals have to establish a good relationship with their patients. This just wasn't as common an experience for African American patients as it was for White patients.

Another important finding of our study was that when patients saw physicians of their same race, in so-called race-concordant visits, patients rated their physicians as more participatory. This would be important for my subsequent work. Specifically, African American patients had less participatory visits with White physicians than White patients did. Asian and Latino, but not White, patients had less participatory visits with African American physicians than African American patients did.

In the discussion of our results, we proposed several reasons that might account for these barriers to partnership and effective communication. First, physicians may unintentionally incorporate racial biases, such as racial and ethnic stereotypes, into their interpretation of patients' symptoms, behaviors, and decision-making. Second, physicians may lack understanding of patients' ethnic and cultural disease models or attributions of symptoms. Third, physicians may be unaware of how their patients' expectations differ from their own. In addition, patient factors, such as low health literacy and educational status, lack of preparation for office visits, lack of self-efficacy regarding managing their own health, and language barriers, could contribute to less participatory visits. I would have to do more work to find out which of these explanations best accounted for our findings.

THE QUALITY OF CLINICAL COMMUNICATION

Our *JAMA* study relied on self-reports by patients, and some people questioned the extent to which patient perceptions might be aligned with actual behaviors by physicians. So I decided to use objective measures of the actual communication that takes place during a clinical encounter.

My colleagues and I did studies in which we recorded routine medical visits, with the consent of patients and their physicians, and then used the Roter Interaction Analysis System to analyze physician-patient communication. Developed by one of my mentors, Debra Roter, the coding system categorizes the medical visit discussion according to four clinical functions: data gathering, educating and counseling patients, relationship building, and partnering with patients to negotiate diagnostic and treatment decisions. For instance, when a physician asks a patient "How have you been feeling over the past month?" or when a patient tells a doctor "I was in the hospital last year for ulcers," those statements are coded as data gathering. Education and counseling might be when a doctor says, "Getting enough sleep is important for your health. Try to go to bed around the same time every night." Relationship building might include statements of empathy or concern, such as "This must be very hard for you" or "I hope you'll be feeling better soon." And partnership-building statements describe when the physician checks to make sure a patient understands or elicits patients' opinions, "How does that sound to you?" We also gave both patients and physicians post-visit surveys to complete in which we asked them to rate the quality of the communication in the visit as well as other attributes of each other.

One of Dr. Roter's earlier studies had shown that physicians tended to focus more on strictly medical topics with Black patients, whereas they had a more balanced content of medical and social topics when speaking with White patients. As an experienced health communication researcher, Dr. Roter guided the team in furthering our understanding of this issue through a more nuanced exploration of the communication behaviors. We decided to analyze a few other measures of communication. One was verbal dominance, which describes the ratio of physician talking to patient talking during a visit. Another was patient-centeredness, which describes the psychosocial, emotional, and partnership-building talk that occurs during a visit. We examined "physician positive affect," which describes the degree of a doctor's assertiveness, interest, responsiveness, and empathy in relation to the patient, as well as the degree to which the doctor sounds rushed. We decided that patient positive affect, which describes the level of assertiveness, interest, friendliness, responsiveness, and empathy in the patient's tone and responses, was similarly important, so we examined that as well. Finally, we measured the duration of a visit and the speed of speaking by both the physician and the patient as communication measures that could indicate partnership and have a relationship to health outcomes.

In listening to the audiotapes, trained coders were also asked to rate the overall emotional tone of the visit for patients and physicians. The two coders who conducted all the ratings were blinded to the study hypotheses and were experienced in using the Roter Interaction Analysis System. In addition, the coders were not informed of the race of the patients or physicians, though they may have been able to guess this information. The coders considered the assertiveness, interest, responsiveness, and empathy of both the patients and the physicians. Coder agreement within one point on the patient and physician positive affect scales ranged from 88 percent to 100 percent.

Our results from coding these audiotapes showed that physicians communicate differently with Black and White patients.[3] For example, physicians talked 73 percent more than Black patients but only 50 percent more than White patients. In other words, physicians are verbally dominant with all patients, but even more so with Black patients than with White patients. Independent coders rated doctors as sounding less friendly, enthusiastic, and responsive with their Black patients. They also rated the Black patients themselves as sounding less friendly, enthusiastic, and responsive. Rachel Johnson, who was an MD/PhD student on our team and my first mentee, led this analysis. In a later study, led by another mentee, we discovered that positive physician affect, or the friendliness of the physician's tone, was a significant predictor of high patient trust, especially among African American patients.[4]

In analyzing our recordings, we also found that the content of the conversations was different between racial and ethnic groups. With African American patients, conversations focused more on medical issues and less on psychosocial matters like family and work, even though these might be contributing to a patient's health and ability to follow through on the physician's recommendations. Physicians conveyed less interest in African American patients' emotional well-being, discussed fewer personal issues, and showed less empathy and concern for these patients, corroborating what the patients were reporting. We called this communication pattern less "patient-centered."

Overall, in comparison to Whites, in studies led by three other mentees, Crystal Cené, a postdoctoral fellow in general internal medicine, Anika Hines, a doctoral student in public health, and Bri Ghods, an MPH/MBA candidate, we found that African Americans with hypertension and depression experienced less participatory communication, more technical and biomedical conversation, and less rapport-building and psychosocial conversation. This was true regardless of the race of the physician.[5]

CONCORDANCE AND DISCORDANCE

I was intrigued by the finding that ethnic minority patients experienced poorer communication overall with their physicians. Because of shared experiences in society, I would have expected these patients to have better communication with physicians with whom they shared the same race or ethnicity—in other words, in race-concordant relationships—than they did with physicians with whom they did not have social or cultural ties—in race-discordant relationships. This could also explain some of the findings from our earlier study, where patients in race-concordant visits rated their physicians as more participatory.

We analyzed the results according to whether patients and physicians shared racial or ethnic backgrounds.[6] Our results showed that race-concordant visits were longer by about 2.5 minutes, patient positive affect was greater, there was more participatory decision-making, and patients reported higher levels of satisfaction. In race-concordant visits, physicians spoke more slowly, patients exhibited a more positive emotional tone, and patients were more likely to recommend the physician to a friend. Overall, patients experienced poorer communication with physicians and, again, reported less participation in decisions about their care in race-discordant visits than in race-concordant visits. This was true for both African American and White patients. We didn't have enough Latino or Asian patients in our study to look at race concordance for these groups.

These studies led to some controversy. My colleagues and I felt that we didn't know which factors were playing the most important roles, so the only clear recommendation we could offer at that time to address poorer experiences among African American patients seeing doctors of another race was to increase racial diversity among physicians. Many in our field felt this was a provocative assertion, and for many years I had to defend this recommendation. I explained that we weren't suggesting that every patient needed to have a physician of the same race. Rather, we were observing that diversifying the physician workforce would provide African American patients with more opportunities to choose a physician of their same race if they felt it was important to their care.

Differences in communication did not fully account for why patients in race-discordant relationships rated their care as worse. We wondered whether other factors, such as physician and patient attitudes (including bias, mistrust, and cultural misunderstanding), might play a role. But before we investigated more of those issues, we wanted to try to unpack the idea of concordance a little more. Patients and physicians can be concordant or discordant in other demographic factors besides race and ethnicity, including gender, age, education, social class, or language. This prompted another study, led by my mentee Rachel Johnson (now Thornton), in which we looked at concordance across a variety of different social dimensions, including race, gender, age, and education.[7]

Data sets for this study were drawn from our two earlier observational studies conducted from July 1998 to June 1999 and January to November 2002 and involved 64 primary care physicians and 489 of their patients from the Baltimore,

Washington, DC, and Northern Virginia areas. Physicians were recruited from group practices and federally qualified health centers, and both studies targeted practices with a high percentage of African American physicians and patients. We obtained consent from all patients and physicians to record their visits. We analyzed the data sets to determine the association of patient-provider social concordance with medical visit communication and patients' perceptions of care. Our analyses revealed that lower patient-physician social concordance was associated with less positive patient perceptions of care and lower positive patient affect. Furthermore, we found that social discordance across multiple characteristics had cumulative negative effects on patient-physician communication and perceptions of care.

In 2008, I worked on another study with communication researchers at Baylor College of Medicine and Texas A & M University to examine the extent to which patients perceived themselves to be similar to their physicians with regard to beliefs and values (as well as race, ethnicity, and culture). We found that, as with race and social concordance, perceived personal similarity was associated with patient-centered communication, trust, and satisfaction.[8] As a result of these studies, we concluded that research should move beyond one-dimensional measures of patient-physician concordance to understand how multiple personal and social characteristics influence health care quality.

A PERFECT MATCH?

These results weren't especially surprising. But they were worrisome, since most minority patients cannot choose a physician of their same racial or ethnic group, due to the underrepresentation of African Americans, Latinos, and Native Americans among physicians. In many health care plans, no one has this option.

Despite the potential benefits of greater concordance between patients and physicians, we know that universal concordance in the United States is not possible—nor should that be the goal. Although diversifying the physician workforce clearly has other advantages, we also need to strive to educate providers to relate to patients of all backgrounds. Medical curricula need to be modified to give physicians a better appreciation not only of the cultural background of their patients but also of historical and structural factors shaping patients' experiences—neighborhood circumstances, access to employment and education, realities of living conditions and job demands, and exposures to discrimination in society. As our society becomes increasingly diverse, all physicians need to be able to communicate effectively with all their patients.

Our findings began to have an influence on policy. Two reports from the Institute of Medicine, *Crossing the Quality Chasm* (2001) and *Unequal Treatment: Confronting Racial and Ethnic Disparities in Health Care* (2003), cited our studies as part of their rationale for national recommendations for health system interventions to increase patient involvement in care, cultural competency training for health professionals, and diversification of the health professions.[9] The Liaison Committee on Medical Education required medical schools to incorporate cultural competency training among their standards. I participated on an advisory group for the American Board of Internal Medicine that developed a recertification module on care of the underserved for internal medicine physicians.

Enhancing physicians' knowledge of cultural, historical, and structural factors impacting the health of patients of color continues to be of great importance. However, many of the physicians I work with and who were included in my studies were disturbed to discover that they were communicating differently with their patients of color, despite their best intentions. I also wanted to study this more, because I didn't think most physicians were doing this intentionally; there had to be other reasons.

RACIAL BIAS

As a result of my studies on the influences of patient race and patient-physician race concordance on clinical communication, in the fall of 2004 I was invited by one of my role models in general internal medicine, Thomas Inui, and my mentor, Debra Roter, to be part of an initiative supported by the Fetzer Foundation focused broadly on the role of relationships in health care. This group, called the Relationship-Centered Care Research Network, worked to describe a new framework for relationships in health care that moved beyond patient-centered care.[10]

Whereas patient-centered care focuses on meeting patients' needs, with the primary relationship of interest being the patient-physician relationship, relationship-centered care applies to all participants in the health care setting and can be looked at from each person's perspective. From a physician's perspective, there are four key relationships. The first is the physician's relationship with self: self-care, beliefs, and personal histories greatly shape the type of care that doctors can provide to their patients. The second is a physician's relationships with his or her colleagues, including clinical teammates, research and educational collaborators, mentors, and trainees. The third is a physician's relationship with patients, which helps determine the quality of care a patient receives. The fourth is a physician's relationship with the community, including their opportunities to build partnerships to support a healthier and more just society.

In addition to these relationships, four basic principles underlie relationship-centered care. First, relationships in health care should reflect and respect the personhood of all participants. Second, emotions and their expression are important components of these relationships. Third, all relationships occur in the context of reciprocal influence. Fourth, the formation and maintenance of genuine relationships in health care is morally valuable.

In our research network, we started to talk about how some factors that had not been studied, such as physician attitudes about a variety of factors, including race, and their emotional responsiveness to patients' nonverbal behaviors, might influence their relationships with patients. I decided to conduct some studies specifically examining racial attitudes to see whether these could explain the adverse outcomes we were seeing.

Racial attitudes are the result of two cognitive processing systems. The first is one's conscious system, which involves the controlled and deliberate processing of stimuli that produce explicit beliefs and attitudes. The second is one's subconscious or unconscious system, which includes automatic responses and skills that are learned through early socialization and repeated exposures. These lead to implicit beliefs, attitudes, and biases that often affect our understanding, actions, and decisions in an unconscious manner.

When we began to consider the impact of implicit biases on health outcomes, we found a number of studies that had examined *explicit* bias among physicians. One such study found that many physicians perceived Blacks more negatively than Whites on intelligence, education, and likelihood of adherence.[11] Another study found that negative beliefs about a patient's race influenced recommendations for coronary artery bypass surgery.[12] Physicians' perceptions of patients' education and physical activity preferences were also significant predictors of their recommendations, independent of clinical factors (such as the severity of their symptoms), appropriateness (such as the anatomical presence of a severe blockage of the left main coronary artery), payer (whether they had private or public health insurance), and physician characteristics. This study further highlighted the need to better understand the ways in which providers' beliefs about patients might explain disparities in treatment.

Neither of these studies looked at the role of *implicit* bias on physician behaviors. Around this time, I learned about a test called the Race Implicit Association Test (IAT), developed by Harvard University researchers, that measures the extent to which people associate the faces of Black and White people with good and bad words. Images appear rapidly on a computer screen and subjects respond by sorting pairs of images and attributes using the right and left keys. The premise is that individuals will respond faster to concepts that are strongly associated compared to those that have weak associations. If subjects match White+good/Black+bad pairings faster than Black+good/White+bad pairings, then the Race IAT score differs from zero and the result is positive. The Race IAT is a measure of an individual's implicit or unconscious attitudes.

The hundreds of thousands of people who have done the test online have revealed that most Americans—and most people from other countries as well—implicitly favor Whites over

Blacks. In a sample of 732,881 respondents on Project Implicit websites between 2000 and 2006, 70 percent of respondents had a pro-White bias on the Race IAT. Results from the website show that members of socially marginalized groups, including Blacks and other people of color, have more positive implicit attitudes toward their own groups than people outside of them, but that they, too, have a moderate preference for the more socially valued group. The developers of the IAT believe that members of marginalized groups develop negative associations about their own group from the broader culture in society, but they also have positive associations because of their own group membership and that of others with whom they associate.

To figure out whether implicit biases were contributing to racial disparities in health care, I reached out to Anthony Greenwald, a social psychologist at the University of Washington in Seattle, who was one of the developers of the IAT. He was enthusiastic about extending some of the work he was already doing where he was measuring implicit attitudes about race in physicians and asked if he could include one of his colleagues, Janice Sabin, in our discussions. I invited them to join our team, which included Dr. Roter, Dr. Inui, Mary Catherine Beach (a general internist colleague and member of our relationship-centered care research network), Kit Carson, a biostatistician who had worked with me for several years, and me. We conducted a study in which we administered the Race IAT to doctors and then measured their communication with African American and

White patients.[13] Our study drew from 40 primary care clinicians (48 percent White, 22 percent Black, 30 percent Asian or from the Indian subcontinent) and 269 patients (79 percent Black, 21 percent White) in urban community-based settings. Clinicians' implicit racial bias and race and compliance stereotyping were measured with two implicit association tests that were then related to audiotape measures of visit communication and patient ratings. Our study showed that the medical world is much like the rest of society: about two-thirds of primary care doctors implicitly preferred Whites over Blacks, matching White faces with good words and Black faces with bad words faster than Black faces with good words and White faces with bad words. In addition, two-thirds of the physicians in the study held the implicit stereotype of Whites as more cooperative and Blacks as more mistrusting or reluctant to cooperate.

We also found that the greater a doctor's unconscious racial bias or stereotype, the more they dominated conversations with Black patients and the less patient-centered they were—that is, the less they discussed social and personal issues—with these patients. At the same time, Black patients perceived these physicians as less respectful and less trustworthy. In our study, as physicians' IAT score increased, indicating a stronger pro-White bias, the odds declined that Black patients would report being satisfied with their visit, being respected by the doctor, being liked by the doctor, liking the doctor, trusting the doctor, and recommending the doctor to a friend. With a few exceptions, pro-White bias had little impact on the visits of White patients; in fact, for some measures, such as respectfulness, it led to more positive ratings.

When we examined these relationships separately for Black and White doctors, we found that the associations of implicit bias or stereotyping with most communication measures were similar in these two groups. However, the associations of implicit bias or stereotyping with patient ratings were not as strong among Black patients seeing Black doctors, suggesting that these patients were more "forgiving" of doctors of their same race. Because this was the first study to explore the links between implicit bias, clinician behaviors, and patient ratings in actual patient encounters, we pointed to the need for additional research to examine the links among implicit race bias and stereotyping and health outcomes.

Throughout my career as a doctor, I had come to notice widely held beliefs among clinicians that they are not biased or that, since they explicitly have positive attitudes toward different groups, implicit biases wouldn't impact what they actually do. However, other studies have substantiated our findings and suggested that these biases could influence clinical decisions and health care outcomes. For example, a study of medical residents found that the greater a doctor's pro-White implicit bias, the lower the intention to treat Black patients suffering from insufficient blood flow to the heart with medications to prevent heart damage—the same medications the doctor would readily provide for White patients in the same condition.[14] Unconscious bias, in other words, causes doctors to make mistakes.

I was humbled to find, when I took the race implicit association test, that I also had a slight implicit preference for Whites over Blacks, even though I had grown up in Africa and felt very proud of my African heritage. I began to pay closer attention to my own behaviors with patients whose social characteristics and life experiences were different from mine, even if we were both Black. It was sobering for me to realize that, even with the best intentions, I too might be guilty of engaging in biased behavior with my Black patients. Some physicians who participated in the study were similarly surprised and concerned by their results.

However, many medical professionals weren't ready to accept that doctors might be directly contributing to racial disparities; our paper on implicit bias took almost four years to get published, after rejections by editor after editor of medical journals, despite positive reviews from peers. It was finally published in 2012 in a special issue of the *American Journal of Public Health* that focused on the science of research on racial/ethnic discrimination and health.

OVERCOMING IMPLICIT BIASES

Understanding what we know about implicit bias is the first step in reducing it. Earlier work suggested that implicit biases are difficult to change. However, some recent studies show that they can be changed by experiences.[15]

The science in this area of health equity research is growing, and many approaches are being tested and applied to change these attitudes or their associated behaviors. One strategy is called stereotype replacement. It involves replacing stereotypical responses with nonstereotypical responses. The first step in stereotype replacement is recognizing that the response you're having is based on a stereotype, not reality. By labeling this response as stereotypical, you can reflect on why you're having this response. That makes it easier to consider how to avoid biased responses and what those responses might look like.

Another strategy to reduce biased behavior is called counter-stereotype imaging. This involves spending time thinking about or imagining people you know who don't conform to a stereotype. The more people we know with different backgrounds, the more opportunities we have to realize the errors of stereotypical thinking.

A third approach is known as individuation. This involves treating people not according to their membership in a category, such as their race or gender, but according to that person's unique traits and characteristics—really focusing on the person's specific strengths, likes and dislikes, and life experiences.

A fourth approach, known as perspective taking, involves putting yourself in the shoes of a member of a stereotyped group. For example, how would it feel to be told that you couldn't apply for a leadership role because of your race or gender? Or to be questioned about your use of illicit drugs when you have a life-threatening condition associated with severe pain, simply because you're a person of color? Perspective-taking increases the degree of psychological closeness to a group and lessens the impact of automatic and stereotypical evaluations.

We can't easily change implicit biases, but one way to interrupt the path between automatic and subconscious beliefs and the behaviors that follow in rapid sequence, if unchecked, is to use a checklist when interacting with people we don't know, especially those whose backgrounds differ from our own. Checklists can help ensure that everyone receives a certain standard of care and prevent misdiagnoses (by clinicians) or mistreatment based on bias—or simple human error.

I've devised a checklist for implicit bias that incorporates what we know about effective physician-patient communication behaviors to drive behavioral change.[16] I don't know if such a checklist would actually work if used, or if many physicians would even be willing to use it, because specific research on the use of checklists to reduce biased behaviors has not been done. However, in addition to other research on the benefits of checklists in general, social psychology research shows that changed behaviors can lead to changed attitudes over time, so this approach holds some promise. My checklist follows the mnemonic RELATE:

R*espect.* Respect the humanity of the person in front of you, regardless of whether you like them or agree with what they're saying;

E*mpathize.* Imagine yourself in the person's shoes;

L*isten.* Listen more and talk less;

A*sk.* Ask yourself what assumptions you're making and whether they're based on facts about this particular person;

T*alk.* Talk with people about their personal lives, and get to know them as individuals;

E*ngage.* Engage people in problem-solving and decision-making by asking their opinions about any joint activities you're considering.

FROM DESCRIPTIONS AND EXPLANATIONS TO SOLUTIONS

My research up to this point had been descriptive and explanatory, but increasingly the findings from my studies made me realize the importance of providing solutions. I couldn't just keep documenting these problems without trying to do something about them. I was ready to move beyond describing the problem of health care disparities and understanding some of its mechanisms to developing interventions and solutions.

CHAPTER 3

Developing Solutions to Health Disparities

WHEN I BEGAN TO STUDY African American patients in primary care settings, I noticed that studies testing potential solutions to common medical problems often excluded people of color. This was partly because these studies often didn't take place in real-world settings such as clinics, churches, senior centers, or other places in the community. Instead, most of the studies involved people who were relatively healthy volunteers and who were willing to drive to research facilities and follow strict treatment protocols. These studies required participants to have high levels of motivation and resources. In other cases, the studies weren't based on information collected from active participants, but rather mined from enormous existing data sets from insurance companies and national surveys.

In contrast, the work I was doing, and the work that I was interested in doing, was taking place in community health centers and clinics that didn't have many resources or abundant staffing and where patients were often struggling with many different medical conditions. For the past 40 years, the Health

Resources and Services Administration (HRSA), an agency of the US Department of Health and Human Services, has supported such community health centers as a source of comprehensive primary health care to people of all ages, races, and ethnicities. The centers even provide care to those without health insurance, with fees based on a person's ability to pay. Some centers specialize in certain populations, such as migrant and seasonal farm workers, individuals and families experiencing homelessness, people living in public housing, or particular minority communities. They're located in medically underserved areas or serve medically underserved populations. They're governed by a community board that understands the needs of the community, and they provide culturally competent care. Patients receiving care in these centers and clinics weren't the ones who typically signed up to be in research studies. They often didn't have transportation, couldn't get time off from work to participate in a strict study protocol, or had medical conditions that might disqualify them from a study that required them to be relatively healthy.

In the studies my colleagues and I began to design, we didn't expect the patients getting their health care in community-based clinics to follow strict regimens that were vastly different from their usual behaviors—regarding, for example, diet and exercise. If we wanted them to undergo a detailed research interview or examination, we provided a transportation voucher and tried to minimize the inconvenience to them. Over time, we would try to design studies so they wouldn't have to travel to locations other than their regular doctor's office, their job, or their home. We worked with people where they were—and this held true for the providers, clinics, and health care administrators as well, not just for the patients.

I also realized while planning studies to test potential solutions that I could apply the findings from my studies of depression to other medical conditions such as hypertension. The issues I had observed regarding the need for information and education, and the concern about communicating with health professionals, were not unique to African American patients with depression. Additionally, we found that many people didn't relate to medical terms such as *depression* or *hypertension*. Instead, they used the terms *emotional health* and *high blood pressure*. As a result of this finding, we adjusted the terminology used in our study materials to promote willingness to participate and acceptance of the programs being tested.

As we'd found for depression, which most African Americans saw as a spiritual rather than a medical condition, we found that for the most part, people believed that hypertension was something that could be felt and that was most likely related to stress. The subjective nature of both these conditions made them difficult to explain, but we wanted to help people understand that depression could be due to a chemical imbalance in their bodies that either contributed to or resulted from negative thoughts and feelings. We also wanted them to understand that, even at times when they might feel well, their blood pressure might still be elevated.

Recognizing widely held beliefs and medical misconceptions within the community enabled us to incorporate people's perspectives into our messages to make them more relatable and effective, using what we called cultural targeting to focus some of our efforts on dispelling myths. Finally, the importance of relationships and communication in my earlier work made it clear to me that developing solutions targeted at improving patient-centered care had great potential to reduce health disparities, and ultimately, improve the physical and mental health of millions of underserved patients.

ADDRESSING DISPARITIES IN HYPERTENSION CARE: FROM FUNDING TO FINDINGS

In the first intervention study I led, our team wanted to explore several questions. Could communication be improved through an intervention aimed at increasing both physician participatory communication skills and patient activation behaviors? Which of these approaches, alone or in combination, would be most effective at improving health outcomes—in this case, hypertension control?

Several studies had shown that interventions that enhance patient activation and participation in decision-making improve patient knowledge, recall of information, adherence to treatment, satisfaction, and clinical outcomes. Yet many of these studies had only small numbers of African American participants, and few had tried to work with both physicians and patients. None of these studies used an approach that took into account what was known about racial disparities in communication or incorporated some degree of cultural and social adaptation.

An adaptation that held a lot of promise was the use of community health workers (CHW) who represent the evolution of community health advisors deployed in the United States in the 1970s to link people in ethnic minority and other underserved communities with resources and health education.[1] CHWs are frontline public health workers who are also trusted members of their communities with an insider's understanding of its people and problems, resources and limitations, and previous solutions that failed.[2] A community health worker builds both individual and community capacity by increasing their neighbors' access to health care along with their health knowledge and self-sufficiency through a range of activities including community health workshops, home visits, informal counseling, social support, and advocacy.

I was ready to test such an intervention that incorporated cultural and social adaptations to address the unique needs of African Americans and persons with low income. So, I began to look for funding opportunities to do that. My prior studies, in which we documented that physicians were less patient centered and engaged African-American patients in decision-making at lower levels than White patients, helped to establish me as an influential researcher and to demonstrate the need for interventions targeting patient-physician communication as a way to address disparities in hypertension control. In 2000, I contributed to a grant application to the National Institutes of Health (NIH) focused on health disparities in which my proposed study was included. Unfortunately, we were denied funding; I was devastated. By this time, other researchers in the field were beginning to cite my findings, and some who did receive funding were asking me to serve as a consultant to *their* grants. I was really worried that others would identify promising interventions long before I had a chance to test my ideas, and that I was going to get beaten to the punch. Fortunately, with support from my mentors, Debra Roter, Neil Powe, Daniel Ford, and David Levine, I recovered from that disappointment and learned that I was going to need to become resilient in the face of rejection if I planned to become a successful researcher.

Later that year, another opportunity to apply for funding surfaced. After months of meeting with the team and the leaders of potential clinical sites, revising our earlier proposal, and enduring countless sleepless nights (culminating in a marathon week in March 2001 that ended with my driving to the NIH campus in Bethesda, Maryland, on the last day to deliver the grant in person), we finally submitted our application. In June, we received the news that we'd gotten an excellent score from the peer reviewers. Later that summer it was confirmed that we would be funded to begin the study in September 2001. I breathed a deep sigh of relief, and then panic set in—what had I gotten myself into? I had never led a large, multimillion-dollar clinical trial before! There was going to be a lot to learn—about recruiting patients, training and managing staff, running team meetings, designing data collection and intervention protocols, and meeting regulatory requirements. Fortunately, my team included experienced researchers, including Drs. Levine and Roter, who were behavioral scientists, and Martha Hill and Lee Bone, who were experienced nurse scientists with expertise in community-based interventions using community health workers to control hypertension.

Between 2001 and 2005, I led a randomized controlled trial called the Triple P Study that focused on patient-physician partnerships to improve adherence to high blood pressure treatment. It involved 41 primary care doctors (23 percent Black, 27 percent Asian, 45 percent White, 5 percent Latino) and 279 patients with high blood pressure, 63 percent of whom were African American and 35 percent of whom were White, from 14 community-based clinics in the Baltimore-Washington, DC, area. An experimental group of doctors received computer-based communication skills training, while the experimental group of patients received in-person visits and five telephone sessions from community health workers who "coached" them to participate more actively in their care.

The community health workers were an African American woman and a Latino man who lived in neighborhoods near where many of the patients lived and were familiar with many of the social, economic, and medical issues they faced. The coaches used questions to elicit patients' concerns and followed that up with skills practice to help them prepare for the visit. Topics ranged from general health concerns to specific concerns related to hypertension knowledge and treatment, lifestyle changes, stress, and other considerations. For example, the coach would ask the patient, "What concerns do you have today about your health or about your high blood pressure?" Then the coach would follow up with "Tell me how easy or difficult it would be to talk with your doctor about this." To encourage skills practice, the coach would say "What would be a good question to ask your doctor?" and then "That's good. Let's write the question in your diary so you'll remember to ask him." One of the tools they provided was pocket-sized diaries for patients to record their appointments, medications, and questions—simple, but effective. The study sought to compare the relative effectiveness of these patient and physician interventions, both separately and in combination with one another, against the effectiveness of minimal interventions. It evaluated the following outcomes measured at enrollment, three months, and twelve months: (1) patient adherence to medication and lifestyle recommendations; (2) patient and physician ratings of quality of care; (3) patient-physician communication behaviors; and (4) health outcomes, including blood pressure control.

The physician communication skills program provided physicians with personalized feedback based on their videotaped performance with a simulated patient during an office appointment. The patient was an African American man with hypertension scripted to present common social barriers, cultural beliefs, and expectations related to adherence with hypertension therapy. The feedback focused on communication skills relevant to increasing patient engagement, activation, and empowerment and was organized within four functions of the medical interview—data-gathering, patient education and counseling, rapport-building, and facilitation and patient activation.[3] In addition, five specific behaviors linked with successful hypertension management were targeted: (1) elicit the full spectrum of patients' concerns; (2) probe patients' hypertension knowledge and beliefs; (3) monitor adherence and identify barriers; (4) assess adherence-related lifestyle and psychosocial issues; and (5) elicit commitment to a therapeutic plan.

Intervention group physicians reviewed the videotape of their personal interviews with the simulated patient and completed exercises on a CD-ROM or in the workbook. Control group physicians participated in the simulated visit but did not receive any feedback until the end of the study. All physicians received a copy of the currently available hypertension treatment guidelines at the beginning of the study and a monthly newsletter with study updates and summaries of recent journal articles.

The patient intervention was based on a pre-visit coaching model shown to improve patients' communication with clinicians and health outcomes.[4] Telephone follow-ups reinforced the importance of preparing for clinic visits with a listing of concerns. Intervention patients also received bimonthly "photonovels"—publications containing a story told in a sequence of photographs with dialogue added in superimposed speech balloons (sort of a researcher graphic novel tool)—that reinforced the coaching messages. All patients in the intervention and control groups received a monthly health newsletter by mail, designed to meet the needs of adult readers with lower literacy.

For data gathering, we were looking for physicians to use open-ended questions, such as "What concerns you the most about your hypertension?" to probe the patient's concerns. On the patients' side, we were looking to see if they would tell their stories and disclose their concerns to the physician; this would indicate patient engagement. For educating and counseling, we were looking for physicians to provide concise, clear information and for patients to tell physicians what they understood and intended to do; this would indicate patient activation. For participation, we looked for physicians to engage patients in problem-solving and decision-making and for patients to ask questions, express their opinions, and state their preferences; this would also indicate patient activation. Finally, for rapport-building, we were looking for physicians to show empathy and offer emotional support to their patients and for patients to openly share their concerns and feelings and ask for the help they needed to overcome barriers; this would indicate patient empowerment.

The combined physician and patient interventions were effective at improving information exchange, participatory decision-making, and systolic blood pressure, particularly among patients whose blood pressure had not been controlled by medication or other means, over 12 months.[5] Patients in the combined interventions group (who had a community health worker coach and whose doctors had received communication skills training) were more satisfied with their clinical visits than those who were in the minimal intervention group or the separate interventions groups. An analysis led by one of my mentees, Chidinma Ibe, for her doctoral thesis found that regardless of whether their physician received communication skills training, for patients who had a community health worker coach, more topics discussed during the coaching session led to patients asking more psychosocial-related questions during their doctor visits, and longer coaching sessions led to more use of engagement strategies by patients to facilitate their participation in visits with their doctors.[6]

We faced several challenges in conducting the study. Only half the physicians we invited agreed to participate; the others cited lack of time or interest as their reason for declining. Another challenge was the high number of patients who didn't return a year later for follow-up blood pressure readings and interviews with our research team. They had moved, changed doctors, had disconnected telephones, or were experiencing other health, social, or financial barriers to engaging with us. Another potential limitation was that the intervention for physicians was limited to a one-time administration, and for patients to one in-person contact.

We concluded that future efforts needed to incentivize physicians to participate and better support patients' broader needs in order for them to take part in the training and coaching programs. In addition, the optimal "dose" of interventions remained unknown, suggesting the need for further work on patient-physician communication as a way of addressing health disparities. We also suggested that future interventions might be strengthened by including health system–level strategies and further emphasis on patients' social and environmental context. Still, this was one of the earliest studies to provide evidence for the effectiveness of community health workers as "coaches" to activate African American patients with a chronic health condition such as hypertension to participate in their care. Although clinical practice did not change immediately as a result of the study, many national reports and clinical guidelines began to mention the potential value of these approaches in addressing racial disparities in the care of medical conditions. Several other studies in different populations of patients, such as those with diabetes and cancer, began to target patient-physician communication as a way to address disparities. My colleagues and I also used the community health worker coaching protocol as the basis for several of our later interventions delivered by community health workers.

ADDRESSING DISPARITIES IN DEPRESSION CARE: FROM STANDARD TO CUSTOMIZED CARE

The second randomized trial I led aimed to address disparities in depression care. It was called Blacks Receiving Interventions for Depression and Gaining Empowerment, or the BRIDGE Study. This was designed to compare the effectiveness of two programs for African American patients with major depressive disorders: one involving standard collaborative care, the other involving patient-centered collaborative care (with cultural targeting). Collaborative care is a model of care delivery in which primary care providers, care managers, or other health care professionals and consultants work together as a team to provide care and monitor patients' progress. Programs using this model were beginning to show improved clinical outcomes and reduced costs for various health conditions, in diverse settings, using different payment mechanisms. However, few of these models had been adapted to the specific needs and concerns of African American patients with depression. The study involved 27 primary care clinicians and 132 African American patients

How to Be an Empowered Patient

People of color, people with less education, and people with modest incomes are typically less active in decision-making with their physicians. They talk less, ask fewer questions, and generally sound less enthusiastic during their medical visits.[1] But many of us, regardless of race, ethnicity, education, or income, prepare less for a doctor's appointment than we should, including not taking the time to research who the best doctor for us might be by asking for referrals or reviewing their professional ratings online. We need to invest the time to find the best fit based on their competence, integrity, compassion, respectfulness, and general approach.

To receive the best care and cures, we all need to arrive ready to actively participate in the conversation in a confident, effective way. Simple things like remembering to take the list of our medications and supplements (don't forget the dosages) and sharing notes on when we first noticed symptoms can make a vital difference. And, honesty is the best policy—we can't allow shame or guilt to get in the way of full disclosures. Remember that all conversations with medical providers are confidential and protected by law, and that your doctor is not there to judge, but to heal.

Today's medical office visits can be fairly brief encounters, so we must make each minute count. Research shows that patients who are more active and effective managers of their own health have more appropriate treatment, better experiences in health care, and better outcomes, including:

- fewer emergency room visits;
- more appropriate use of screening tests for breast cancer;
- lower rates of obesity and smoking; and
- better control of diabetes and cholesterol levels, among other benefits.[2]

Everyone is their own best advocate (and for their family members) in health care, especially if they belong to a socially marginalized group. Here's a quick way of remembering how we can do our "PART" during office visits:[3]

***P**repare.* Arrive with your medications list, and notes about any problems, symptoms or questions that have come up since your (or your family member's) last visit. Be clear about what you (or you and your family member) want to get out of the visit.

***A**ctively engage:*

- Talk with your doctor about anything that may be worrying you (or assist your family member in raising all concerns) by clarifying or expanding medical history and introducing relevant topics as needed. Confirm the accuracy of any advice you've gotten from laypersons.
- Ask questions (or assist your family member by reminding them of questions identified during the visit preparation as needed).
- Indicate your treatment preferences (or assist your family member by asking directly for their opinion and treatment preferences).

- Clarify instructions or explanations (for yourself or your family member) by using teach back and summarization—for example, "This is my understanding of what we discussed—am I correct?"
- Identify any barriers you or your family member are experiencing to managing your (or their) health and suggest ways to address these challenges.

***R**eview key recommendations.* Write down instructions and new information about the treatment plan given during the visit. This will help cement the information in your memory so you're more likely to follow through on any instructions.

***T**ake recommendations home.* Once you return home, make a to-do list (for yourself, or jointly with your family member) from your notes that you can easily follow, and keep it in a visible place.

NOTES

1. Lisa A. Cooper, Mary Catherine Beach, Rachel Lyn Johnson, and Thomas S. Inui. 2006. Delving below the surface: Understanding how race and ethnicity influence relationships in health care. *Journal of General Internal Medicine* 21 Supp. 1: S21–S27.
2. Jessica Greene and Judith H Hibbard. 2012. Why does patient activation matter? An examination of the relationships between patient activation and health-related outcomes. *Journal of General Internal Medicine* 27(5): 520–526.

3. L. Ebony Boulware, Patti L. Ephraim, Felicia Hill-Briggs, Debra L. Roter, Lee R. Bone, Jennifer L. Wolff, LaPricia Lewis-Boyer, David M. Levine, Raquel C. Greer, Deidra Crews, Kimberly A. Gudzune, Michael C. Albert, Hema C. Ramamurthy, Jessica Ameling, Cleomontina A. Davenport, Hui-Jee Lee, Jane F. Pendergast, Nae-Yuh Wang, Kathryn A. Carson, Valerie Sneed, Kimberly Gudzune, Michael Albert, Sarah J. Flynn, Dwyan Monroe, Debra Hickman, Leon Purnell, Michelle Simmons, Annette Fisher, Nicole DePasquale, Jeanne Charleston, Hanan Aboumatar, Ashley N. Cabacungan, and Lisa A Cooper. 2020. Hypertension self-management in socially disadvantaged African Americans: The Achieving Blood Pressure Control Together (ACT) randomized comparative effectiveness trial. *Journal of General Internal Medicine* 35(1):142–152.

from a wide range of educational and economic backgrounds who were receiving care in urban community-based practices in Maryland and Delaware and planned to continue their care at these practices over the next 12 months. This provided a continuity with the participants that the previous study didn't permit.

In the group receiving a standard collaborative care intervention, clinicians participated in a one-on-one, disease-oriented educational session delivered by a primary care physician with special training in depression. This is an approach similar to the one used by pharmaceutical sales forces, called academic detailing. These physicians also had access to psychiatrists for mental health consultations. The patients of these physicians received a standard collaborative care intervention that focused on disease management.

In a second group that received a patient-centered and culturally targeted collaborative care intervention, clinicians completed a case-based, interactive, multimedia CD-ROM skills training program that included simulated casework and individualized feedback regarding their communication with patients (similar to the program used in the Triple P Study, but this time the patient case was an African American woman with depression). Like clinicians in the standard intervention group, these clinicians also received an academic detailing visit. Patients of this second group of clinicians received a patient-centered and culturally tailored intervention, including care management focused on access barriers, the social context, and patient-provider relationships.

In the first patient group, the standard depression care manager was a White woman. In the second group, the patient-centered depression care manager was an African American woman. Both were social workers with clinical experience who provided educational materials, assessed patients' status, and encouraged adherence to recommended treatments. However, in addition to the standard needs assessment questions, the patient-centered assessment explored access barriers, patients' use of spirituality as an active coping strategy, concerns about treatment, social stressors known to disproportionately affect African Americans, and communication problems with health professionals. The depression case manager in the patient-centered group also used individualized approaches. For example, instead of sending these patients generic depression educational materials, she sent them culturally targeted materials designed to address barriers to depression treatment.

We found that both groups of patients showed similar clinical improvements in depression severity and mental health functioning scores.[7] Treatment rates were higher with the standard approach; however, adherence to care management and patient experiences were better with the patient-centered approach. Our results didn't justify advocating strongly for one approach over the other, since there were minimal differences in clinical improvement between the two groups. Rather, we suggested that depending on the populations served and resources available, either program would likely lead to improved health outcomes and should be considered appropriate for treatment of African Americans with depression in primary care settings. For health systems caring for predominantly African American populations, this study provided a program that was just as effective as the standard treatment, and that African Americans would likely be more adherent to and would prefer. Cheryl Jackson, one of the physicians who was in the study, said, "the BRIDGE Study raised my consciousness about how you communicate with and connect with people who are depressed by giving me a structured approach to listening with respect, and examining my own biases."

One of the BRIDGE Study patients, Camille Yarborough (not her real name), was 38 years old and working full time while taking graduate school classes. She was depressed and was feeling overextended but was deeply worried about becoming addicted to antidepressant medications. However, after working with her BRIDGE Study care manager and primary care doctor, she agreed to try medication, and began to sleep better and suffer fewer emotional breaks than before. Her care manager also helped her locate a Christian counseling program, which was important to her. The counseling helped her look at her family dynamics and begin to learn how to set boundaries and "say no." It's always rewarding when these studies not only provide findings for us but actively help patients like Camille in the process, too.

BUILDING CULTURAL AND STRUCTURAL COMPETENCIES AMONG CLINICIANS

Our two intervention studies provided new knowledge about the effectiveness of culturally appropriate patient activation interventions and tailored physician communication training skills programs for reducing racial and ethnic disparities in health care for patients with chronic medical conditions.

However, clinicians need a broad set of professional competencies to deliver equitable care to patients from groups experiencing health disparities. They must have knowledge about established and evolving biomedical, clinical, and related sciences. They then need to be able to apply this knowledge to deliver compassionate, appropriate, and effective care. This requires practice-based learning and patient-centered communication skills, and it also requires professionalism, as evidenced by carrying out these responsibilities and adhering to ethical principles. Clinicians must engage in systems-based practice, which means that they're aware of and responsive to the larger context of health care and use system resources effectively to optimize patient care.

Another set of professional competencies—cultural and linguistic—is important in providing high-quality care to patients who come from cultural groups that differ from those of most health professionals in a society. Effective communication requires, at a fundamental level, communication in a language that the patient understands, including knowing when and how to use an interpreter. For example, clinicians should use professionally trained interpreters whenever possible, or bilingual staff, rather than family members of patients to learn what ails them (to avoid any family dynamics that might color honest presentation of facts), and they should look at patients and speak directly to them rather than to the interpreter. As with race and ethnicity, language concordance may not always be possible. For that reason, all physicians need to be culturally and linguistically competent to improve health equity.

Beyond communication, cultural competence allows a health care provider to understand and respond to the needs a patient brings to a health care encounter. It also enables a clinician to develop effective interpersonal and working relationships that supersede cultural differences. Cultural competence includes being aware of one's own worldview, developing positive attitudes toward cultural differences, and gaining knowledge of different cultural practices and worldviews. Awareness of the cultural values, beliefs, and practices of different racial and ethnic groups can allow health care providers to address the unique risk factors of groups that differ from their own. For instance, some South Asians consume Ayurvedic medicines that may expose them to toxic metals. This knowledge should not be used to automatically assume that all South Asians engage in this behavior, but it could be critical to reach the correct diagnosis if a South Asian person were to present in a clinical setting with metal toxicity.

Cultural competence is being increasingly taught in medical education, but training is not standardized, and little research has been done to assesses whether students have acquired and use the skills they're being taught and whether this training actually improves patient experiences and outcomes.[8] Rather, most studies of cultural competence ask learners to self-report whether they believe their attitudes and skills have improved. Assessing the success of cultural competence training remains vital.

Training in cultural competency also runs the risk of driving reductionist behaviors. The cultural competence framework used in medical education can expose physicians to homogenized and streamlined ideas of "culture" in an effort to help them assess a patient's behavior, preference, or response. This knowledge of cultural customs has benefits, but making decisions about a patient's care based on these assumptions can be inadequate. Such training can end up promoting stereotypes and the development of physician bias, although in different ways than before such training.

Some have proposed *cultural humility* as a better term than *cultural competence*. Cultural humility is the notion that providers can exercise self-awareness to foster respectful relationships with patients. It places the emphasis on aspects of cultural identity that are most important to a patient. This framework acknowledges the complex formation of individual identity and belief. Cultural humility "incorporates a lifelong commitment to self-evaluation and critique, to redressing the power imbalances in the physician-patient dynamic, and to developing mutually beneficial and non-paternalistic partnerships with communities on behalf of individuals and defined populations."[9]

The idea of cultural competence is also related to patient- and relationship-centeredness, discussed in chapters 1 and 2. Each of these approaches is founded on the ability of health care providers and systems to see patients as unique people, to build effective rapport, to explore patients' beliefs and values, and to find common ground regarding treatment plans.[10] They are distinct, however, in that cultural competence calls for health care providers to exhibit other qualities, such as understanding the meaning and importance of culture and effectively using interpreter services. This is important everywhere, but critical in areas with large immigrant or refugee populations.

Just as patient-centered care has evolved to relationship-centered care, cultural competence has evolved to what we now call structural competence. Structural competence amplifies physicians' and health care organizations' responsibilities, extending from the cultural and linguistic needs of the patients and communities they serve to their broader social needs, shaped by structural factors. On the individual clinician or staff level, structural competence implies understanding how forces at the institutional and societal levels drive health inequities and taking these factors into account when managing patients. Much work remains to be done in enhancing structural competence among clinicians.

BUILDING CULTURAL AND STRUCTURAL COMPETENCIES WITHIN HEALTH CARE ORGANIZATIONS

Efforts to advance health equity are increasingly focused not only on the role of individual clinicians and staff, but also on the role of the health care systems themselves. Many experts believe these systems ought to demonstrate a combination of patient/relationship-centeredness, cultural competence, and structural competence to promote health equity. On a system level, overlapping features across patient/relationship-centered care, cultural competence, and structural competence include the general endorsement that services should be aligned to meet patient needs; that health care should be available and accessible to patients (that now also includes telehealth services); that educational materials should be tailored to patients' needs, including literacy, preferred language, and digital access; and that information on performance should be publicly available.

On an organizational level, patient-centered care also calls for patients to be able to get same-day appointments and for maintaining continuity and secure transitions across health care settings. In addition, patient centeredness calls for general information on patient experiences to be tracked, reported, and addressed by these corporations, government agencies, hospitals, institutions, and practices.

System-level cultural and linguistic competence, on the other hand, means the practice or organization should provide patients who have limited English proficiency with access to culturally and linguistically appropriate oral and written language services through such means as bilingual/bicultural staff, trained medical interpreters, and qualified translators. This has become more difficult as language diversity in the United States has grown, but it's vital. System-level cultural competence also includes a diverse workforce that reflects the patient population and partnering with communities in setting priorities and planning. In addition, cultural competence calls for quality measures stratified by race and ethnicity.

Structural competence on a system level is reflected in a health care organization's focus on helping to address broader institutional and societal forces perpetuating health disparities. Health system leaders who demonstrate structural competence recognize the societal forces that shape the interventions they use and are humble about the use of those interventions. Other system-level demonstrations of structural competence include an explicit mission to address health equity; efforts to address patients' broader social needs as part of health care delivery; providing competency-based curricula on structural factors, including structural racism, to organizational members; and appropriate resourcing of health equity efforts.[11] In addition to the measures of accountability described for patient-centeredness and cultural competence, structural competence calls for local community health data to be tracked, reported, and addressed.

One example of cultural and structural competence can be found in an initiative that was launched by the Minnesota Department of Health in 2011. The state started requiring health care providers to collect race, ethnicity, and language data in an effort to take targeted actions to reduce inequities for their minority populations. Based on the data, interventions were developed and implemented that had positive effects on identifying and addressing health disparities.

BUILDING TRUSTWORTHINESS IN HEALTH CARE SYSTEMS

Just as trustworthiness is critical in patient-physician relationships, the ethical behavior and reliability of organizations has become a focus of health equity efforts in recent years. Relationships at multiple levels can thrive only when a health care system provides the support for that to happen. This requires leadership, patient- and relationship-centered practices and policies, and cultural and structural competence. As described in chapter 2, trustworthiness is achieved through demonstration of benevolence, integrity, and competence, all of which are achieved through transparency. A clear commitment to health equity, open information systems for communication and reporting, comprehensive needs assessments to inform care, transparency, and collaborative partnerships can all engender trust in health care systems by underserved communities.[12]

In 2019, Donald Wesson, Catherine Lucey, and I described how health systems build trust to eliminate health disparities in the *Journal of the American Medical Association*.[13] We acknowledged that health care system leaders often have the standing within the larger community to facilitate collaborations. However, achieving the trustworthiness to be effective in this action requires that these leaders acknowledge the tensions that disadvantaged groups feel toward the health care system and the reasons for those tensions. Leaders need to demonstrate respect for the assets that exist in communities and leverage these assets to address health disparities. They must communicate transparently, authentically, and frequently with partners to create strong and lasting trust.

On a systems level, many strategies can build trust between the health care system and underserved communities. These include:

- continuously seeking, developing, and nurturing trust-based relationships with community institutions;
- establishing institutional commitments with appropriate operational strategies, resources, and accountability systems;
- adopting co-production models that engage and empower community institutions to work as co-equals

in the identification and design of interventions and dissemination of results;

- establishing, monitoring, and sharing progress on metrics that measure progress toward agreed-on areas of focus; and
- establishing supporting systems for, and measuring compliance with, an institutional commitment that all interactions with the community undertaken by the institution are conducted in alignment with respectful practices for community engagement.

MOVING BEYOND HEALTH CARE SYSTEMS

Up to this point, my intervention studies focused mainly on patient-physician relationships and what happened in examination rooms within clinics. However, my research team also led several systematic reviews of the existing research on individual clinician and system-level approaches to health equity. We identified several gaps in existing research. These gaps included the lack of adequate comparison groups; use of subjective rather than objective measures of individual behaviors and organizational practices; and the failure to measure health outcomes and equity of services across racial and ethnic groups. Through this research and clinical work, we were beginning to see how important it is to study these factors—not only in examination rooms, but also in clinical team and health system leadership meetings. Patients and health professionals bring into their relationship all the other aspects of their life experiences. These experiences are shaped by other relationships, including those with their family members, friends, and co-workers. They're also shaped by factors in the environment, including their neighborhoods and communities, the various organizations in which they participate, and societal values, norms, and policies.

I was now ready to broaden the scope of my research beyond the patient-physician relationship to other relationships and to factors inside and outside of health care that could improve health outcomes and advance health equity on a local community level as well as nationally and globally.

CHAPTER 4

The Johns Hopkins Center for Health Equity

THROUGH THE FIRST 15 YEARS OF MY CAREER, I focused my research on the role of patient-physician relationships in health disparities, the impacts of race concordance between physicians and patients on the treatment and outcomes of primary care patients suffering from depression and hypertension, and the impacts of clinician and, to some extent, health system competencies on health equity. It was this work (including the Triple P and BRIDGE studies mentioned in the last chapter) that caught the attention of the MacArthur Foundation, leading to my being named a MacArthur Fellow in 2007. The MacArthur Fellowship is a five-year financial award given to individuals whom the Foundation identifies as exceptional—often pioneers in their fields—who will demonstrably contribute to the public good. Since the fellowship, unlike most major grant awards, has no specific obligations or reporting requirements, it provides recipients with the flexibility to pursue novel artistic, intellectual, and professional activities. Beyond the tremendous honor, this experience opened doors for me and allowed my career to flourish in ways I'd never imagined.

Becoming a MacArthur Fellow brought everything full circle. I've been inspired and energized by the innovative work of other fellows, including Paul Farmer's remarkable health care delivery, research, and advocacy activity to address global health inequities through Partners in Health, Brian Stevenson's racial justice, criminal justice reform, and public education efforts through his establishment of the Equal Justice Initiative, Latoya Ruby Frazier's photojournalism centered on the intersection of social inequality and health in the postindustrial age, and Nikole Hannah-Jones's investigative journalism covering racial injustice, including her creation of the *New York Times*'s 1619 Project. The fellowship brought attention to my research, expanding my reach to existing and new audiences. Through these interactions, I began to see more clearly how I could connect my experiences as a young African girl who grew up knowing both sides of the disparities problem. I had experienced privilege, as the daughter of a physician in a poor country, and relative disadvantage, as a young woman of African descent in a White, male-dominated profession in America. This helped me bring more empathetic and effective attention to the problems of health disparities here in the United States and around the world. It also inspired me to think bigger—for instance, by partnering with community members who could offer more direct feedback on ways to address their challenges and by designing longer-term solutions. But now I had to figure out how to map my way forward to meet those goals.

In 2010, I was still working at Johns Hopkins as a clinician researcher and sharing the results of my work with national and international audiences, but I was starting to feel the need to take a more holistic approach to the study and application of solutions to inequities. With encouragement from my then division director of General Internal Medicine, Fred Brancati, a brilliant and fearless colleague, I convened a team of about 30 faculty members to create a bold new vision for disparities research. My colleagues and I applied for and were awarded a grant from the National Heart, Lung, and Blood Institute of the National Institutes of Health (NIH) to establish the Johns Hopkins Center for Health Equity, then known as the Johns Hopkins Center to Eliminate Cardiovascular Health Disparities. We became one of ten Centers for Population Health and Health Disparities funded by NIH, five of which focus on cardiovascular health disparities and five of which focus on cancer disparities. This was a multimillion-dollar award as compared with the low six-figure grants we'd previously received.

The Center originally aimed to reduce the incidence and mortality levels of cardiovascular disease and to improve the experiences and outcomes of health care for African Americans and others affected by disparities in Baltimore. The death rate due to cardiovascular disease is astounding among the general population, resulting in 2,200 deaths in America every day, but it's even worse for African Americans, with mortality rates 1.5 times higher for African American men than White men.

Uncontrolled hypertension is the primary contributor to the morbidity (for example, heart attacks, congestive heart failure, strokes, kidney disease) and mortality rate disparities in cardiovascular disease between Blacks and Whites. The issue was personal for me since I lost my father to the disease when he was only 56.

Since its inception, the Center has conducted health interventions that bridge several fields of theory and research to target factors that simultaneously contribute to disparities at individual, family, organizational, community, and policy levels. We engage providers, payers, community organization leaders, residents, and policymakers in our intervention planning, implementation, and dissemination efforts.[1] For example, we targeted low levels of dietary intake of potassium, a result of poor access to fruits and vegetables in "food deserts" that prompts a biological mechanism for hypertension among African Americans. We also targeted health professionals' beliefs and behaviors and health care organizational factors (based on the findings of our previous research) by incorporating care managers and community health workers to help primary care practices better identify and address patients' social determinants of health.

We have an interdisciplinary team that includes faculty, staff, and trainees who come primarily from a collaboration of three Johns Hopkins schools: the School of Medicine, the Bloomberg School of Public Health, and the School of Nursing. Over time, our collaborations expanded to include the Carey Business

School, the Krieger School of Arts & Sciences, and the Applied Physics Laboratory. The Center includes integrated and complementary cores or support units for stakeholder engagement, research resources, and education and training. A community advisory board facilitates relationships among faculty researchers, their trainees, and our community partners.

We use implementation science, which emphasizes actionable research that integrates findings and evidence into health care practice and policy. We also employ community-based participatory research to involve all partners in the research process and to make the most of the unique strengths that each brings to the partnership. By directly engaging the community, we're able to see firsthand the kinds of problems that affect our most vulnerable populations and then hear possible solutions from those same people, which is valuable not only for inclusivity but also as a check on any assumptions we may make along the way.

COMMUNITY-BASED PARTICIPATORY RESEARCH

We researchers need to make sure we leave the comfort of our offices and labs to physically go into communities in order to engage with the people who are running their own organizations and learn about the struggles they're having. These organizations can help to convene meetings with legislators, public health officials, secular and religious community representatives, private sector representatives, and others seeking action to end health disparities. These relationships take time—communities need to know that you're committed to them for the long haul, not just for a single project. As with all relationships, we have to earn their trust.

Partnerships with communities are at the core of our work at the Center. In many disadvantaged areas, people feel that their complaints go unheard, that they don't have a voice. We work to convey that we value and respect people everywhere. Community stakeholders are involved in all phases of the Center's research. They help develop grant proposals, pilot and refine research procedures and materials, recruit study participants, train the staff, and coauthor and write research articles. The more we know about what's going on in local organizations, neighborhoods, or groups, the better we're able to target our research—to make it more relevant, appropriate, and effective for those we're trying to help.

Our focus on community relationships is informed by an approach to research known as community-based participatory research (CBPR). Historically, researchers rather than community members have led research agendas and processes. They've selected the health issues to study, developed study designs, recruited study participants, analyzed results, and reported the results in scientific journals for other researchers and health professionals, all in a bit of an academic vacuum.

Community-based participatory research is "a collaborative research approach that is designed to ensure and establish structures for participation by communities affected by the issue being studied, representatives of organizations, and researchers in all aspects of the research process to improve health and well-being through taking action, including social change."[2] It implies shared decision-making, reciprocal relationships, co-learning, trust, and transparency.[3] In CBPR, community members are actively engaged in identifying the health issues that are of greatest importance to them, are involved in the study design and funding of proposals, help in recruiting, develop intervention approaches, and participate in interpreting, disseminating, and translating findings. Community members don't feel that they're at the mercy of a researcher. Instead, they take a more active role as equal partners in developing and implementing interventions to improve the health of their specific communities. It removes the "us versus them" component of problem-solving, and checks the instinct for one-size-fits-all approaches.

Many injustices have unfortunately occurred in Black and other minority communities across the country, leaving their members mistrustful of the medical community. CBPR builds strong ties among researchers, health care provider networks, community members, and policymakers. Our researchers keep community partners informed of all activities through the Center's website, a monthly electronic newsletter, telephone calls

Involving Communities in Solving Disparities

Community-engaged research and practices are vital in reducing health disparities and ensuring progress toward achieving health equity. Historically, policies and programs to improve health outcomes have been designed by academics and other professionals, with minimal input from the target populations that were meant to adopt them. The movement toward community-based participatory research (CBPR) provides active involvement and collaboration on the front end of research in specific areas by involving community members, organizational leaders, and researchers in all aspects of the process. Simply put, it empowers members of the community to be actively involved in identifying and solving their own problems. This engagement and participation can be achieved at all levels, including:

- identifying and advocating for community priorities;
- collaborative design, implementation, and evaluation of research studies (appropriate recruitment and retention methods; successful implementation of interventions; dissemination of research findings);
- mitigating the challenge of translating research findings into long-term practice; and
- influencing policies that reduce health disparities.

CBPR has several additional benefits including:

- improving the quality of academic-community interactions (shared power and decision-making and more trust-based relationships);
- the appropriateness of interventions (acceptable and culturally sensitive approaches; community-specific and structurally relevant interventions that address social determinants of health; sustainability of the interventions); and
- community benefits (capacity building in existing programs; improved health and reduction of disparities).

FURTHER READING

B.A. Israel, A.J. Schulz, E.A. Parker, and A.B. Becker. 1998. Review of community-based research: Assessing partnership approaches to improve public health. *Annual Review of Public Health* 19:173–202.

N. Wallerstein and B. Duran. 2010. Community-based participatory research contributions to intervention research: The intersection of science and practice to improve health equity. *American Journal of Public Health* 100 Suppl. 1: S40–S46.

A. O'Mara-Eves, G. Brunton, S. Oliver, J. Kavanagh, F. Jamal, and J. Thomas. 2015. The effectiveness of community engagement in public health interventions for disadvantaged groups: A meta-analysis. *BMC Public Health* 15: 129.

or emails, and project-related meetings. All these partners share their expertise and assume responsibilities, and the project's results are co-owned, which helps establish that trust. By working together, we've been able to test comprehensive, multilevel interventions that we hope will speed the translation of evidence-based approaches into clinical and public health practice in Baltimore and beyond.

THE CENTER'S COMMUNITY ADVISORY BOARD

This emphasis on community engagement is especially evident in the Center's reliance on our community advisory board.[4] This group includes local residents and leaders who understand the barriers facing communities and can offer practical, specific solutions to overcoming them. They're often people who worked with me or a colleague on a previous project. The board advises each of the Center's five initiatives—research and translation, education and training, community engagement, public policy, and local to global learning.

Many of our board members have experienced the disparities we study firsthand. Michelle Simmons, who joined us in 2011, is a perfect example, having been raised in a low-income household in West Baltimore, where she's lived her entire life. Back in the 1950s, even fewer choices for health care existed; the system was limited and did little to serve her community. Her family had to sometimes choose whether to eat or pay for her medicine, and the food they did eat was not particularly healthy. When her health began to deteriorate, she didn't want to face the fact that she was becoming a diabetic. Other family members had developed heart disease, emphysema, diabetes (one cousin was an amputee who died from it), and cancer. Her mother died of a massive heart attack. "Solutions to these problems cannot just be medical," Mrs. Simmons says. "It's systemic. And that means that everybody has to get involved." She has represented the Center at national meetings and as a patient advisor for an initiative we partnered on with the American Medical Association called Improving Health Outcomes. She's committed to improving not only her own health through her hands-on contributions, but also that of her community, and as the Center's reach increases, the nation. "I'm a fighter. I'm a believer," she said. "I stand up for what's right. And what's right is people's health, and I will never give up as long as I have breath in my body."[5] Mrs. Simmons realizes that even if she improves her own health, if she doesn't also do something to help address the broader social injustices that contribute to poor health in her community, these conditions will continue to have negative effects—sleep disturbances, exposure to violence, among others—on her, her family, and her neighbors.

The Center established its community advisory board in 2010 with the twin goals of ensuring the relevance of the current and future studies to the needs and interests of communities experiencing health disparities, and ensuring rapid transfer of new knowledge into clinical and public health practice and policy. The 50- to 60-member board is composed of representatives from community-based organizations, neighborhood associations, universities, government agencies, and patients and doctors. I'm a joint chair of the board along with a well-known and highly regarded leader of a community-based organization, Reverend Debra Hickman. Reverend Hickman is the co-founder and CEO of Sisters Together and Reaching, Inc. (STAR), a faith-based, nonprofit community organization that provides spiritual support, direct services, and prevention education to African American men and women affected by HIV. STAR has trained community workers to support people in addressing their health and social needs for many years. Because of this, we partnered with them to deploy community health workers as part of the RICH LIFE patient intervention, described later in this chapter.

Early in the history of the board, we sat down and decided which principles should guide our relationships. These principles include respect for diverse perspectives; partnership in decision-making; clear, ongoing, and regular communication; and trustworthiness and transparency. Every two years, we hold strategic planning retreats. Our most recent one was led by my mentee Chidinma Ibe, the Center's associate director for patient and stakeholder engagement efforts. We also conduct an annual review to reaffirm our principles and the board's mission, vision, and goals. The board's mission is to promote health equity in communities locally and globally through strong community-academic partnerships.

It's important that we build relationships with our community that extend beyond our advisory board. My colleagues, staff, and I attend community forums to talk with and hear from community members and to connect people with resources. We also provide nutritional counseling and blood pressure screenings at local health fairs and block parties. We tell people about research methods and results, including how to be involved and contribute to research. We also conduct special training programs, distribute fact sheets, use website and social media posts, and provide newspaper, magazine, radio, and television interviews. Together, with our community advisory board members, we advocate for change with health system leaders and help educate the community about health and research. We work to ensure that the needs of people are being met and that groups of people historically overlooked in conversations about health and health care are included. We've covered topics ranging from how to prevent and manage health conditions such as hypertension, heart disease, diabetes, depression, and kidney disease to how to communicate with health professionals, obtain testing and treatment, and advocate for oneself and one's family members during the COVID-19 pandemic. We also serve as advisors to community organizations and public health and other government agencies and policymakers as needed.

Building a cohesive board that works well together has required good timing, shared values, commitment, and a strong focus on respect and collaboration. As a result of these qualities, our board has a very low rate of membership turnover, which allows our relationships to thrive, enhances our effectiveness in decision-making, and increases our ability to connect each other to appropriate opportunities. Our conversations and collaborations with the community, both through these advisors and through other community partnerships, are essential to our work.

FAITH-BASED HEALTH PROMOTION AND DISEASE PREVENTION

One way that I've developed relationships in my own community is through my church. For me, faith and health have always been deeply intertwined, and I also found this to be true in my studies of African Americans with depression and other ailments. I've delivered sermons that raise awareness of health equity issues, that promote the need for good relationships, and that emphasize the power of communities. In one of my sermons, I told the congregation, "Many years ago, I felt God calling me to demonstrate His holiness in my life by speaking out when I saw patients from poor inner-city neighborhoods being ignored or treated with disrespect by my colleagues, and sadly acknowledged that I was engaging in some negative stereotyping behaviors with these patients myself. I decided to devote my career to bringing attention to disparities in health care." Being involved in my church has made me more aware of the important role that community institutions play in fostering health and well-being, especially for the most vulnerable people in our communities, including the poor, older persons, and those who are sick or living with physical or mental disabilities. Acting as a leader in my congregation has also helped me to increase my immediate community's investment in reducing health disparities.

While there are many invaluable organizations in communities that help researchers and medical professionals work to improve health and address disparities, churches and other spiritual centers have characteristics that make them particularly effective in working toward our goals.

Beyond the healing role of prayer and meditation for individuals of faith, religious institutions throughout history have lived the words of their creeds by building hospitals, sending missions to far-flung communities in need, and tending to their flocks in remote rural towns and major urban centers alike. In many of America's 300,000 religious centers, both spiritual and physical needs are addressed through community outreach and health ministries. These centers of worship are integrated into the broader social structures of their communities. The trust they earn week to week and year to year, and the compassion and inclusivity they represent through relationships across generations, create a distinct opportunity to address community health concerns that align with public health goals. They're essential bridges to better health, particularly among populations where there's a mistrust of government health programs or where misinformation proliferates on diseases and their cures.

African American churches have served as the centers of their communities through centuries of their congregants' disenfranchisement—their lack of access to legal, political, financial, and even medical resources—so their roles in health care are part of a longstanding holistic mission tradition. They're trusted and accessible resources, as many spiritual centers are, for everything from maternity care to addiction assistance to dietary recommendations to free blood pressure and cancer screenings. People in communities of color can feel powerless when struck by debilitating, frightening health conditions, a feeling that's compounded by a lack of access to or inability to afford effective health care, and the knowledge that their community suffers more than others because of such historic and systemic injustices. Faith leaders and volunteer members of health ministries lead people through a variety of means toward individual self-efficacy over an array of ills that impact and concern them, and thus their communities. These can be as simple as offering health literature in their vestibules or mentioning free flu shots in their weekly bulletins, or as complex as devoting a series of sermons to health topics or partnering in research projects organized by academic institutions. Through the work of all these dedicated people, the empowering messages of the beloved gospel hymns "We Shall Overcome" and "Lift Every Voice and Sing" are translated into hands-on practices. In the next chapter I'll speak more about one successful example of this—the FAITH! (Fostering African-American Improvement in Total Health) program based on an American Heart Association dietary and behaviors framework, led by my mentee LaPrincess Brewer of the Mayo Clinic who collaborated with African American churches in the Rochester, Minnesota, area.

Faith-based organizations often need funding, infrastructure, and technical or personnel support to tackle health disparities appropriately. To engage religious community leaders as partners in advancing health equity, in addition to seeking their counsel in the development of programs, partners from health care and other sectors can help these leaders and organizations by investing in tools and providing capacity-building assistance. A number of initiatives and organizations are working toward that end. For example, President George W. Bush's 2001 Office of Faith-Based and Community Initiatives (since renamed the White House Faith and Opportunity Initiative) enables worship centers' access to governmental medical (and other) resources. The Health Ministries Association, an affiliate of the American Nurses Association, established a series of church clinics staffed by a physician, nurse, social worker, and pastoral counselor that led to training programs for parish nurses and then helping to establish standards for faith community nurse roles nationwide. The Johns Hopkins Bayview Medical Center Healthy Community Partnership works with faith-based organizations to provide health education and equip individuals and groups with access, resources, and tools to become stronger advocates for health and medical care. The Wesley Theological Seminary's Heal the Sick Program, where I've served on the board for 10 years, assists faith community members and leaders in linking their ministries with health care and public health institutions and other community-based organizations.

Many other faith-based efforts across the country and around the world contribute to public health through program development and resource allocations, including the Interfaith Health Program at Emory University in Atlanta; the Center for Faith and Community Health Transformation in Chicago; the Methodist LeBonheur Center of Excellence in Faith and Health in Memphis; the National Black Church Initiative's recent efforts in COVID-19 vaccinations; and the Buddhist Tzu Chi ("Compassionate Relief") Medical Foundation in Los Angeles.

Loving one's neighbor isn't just about feelings; it's also about actions. It's about wishing the best for others and working to translate those wishes into realities that will protect the physical and mental health of the whole community. For me, that work includes the action-oriented research on solutions that we conduct at the Center. Here are a few examples.

PROJECT RED CHIP

The first large study that we ran at the Center was called Reducing Disparities and Controlling Hypertension in Primary Care (Project ReD CHiP), which operated from 2010 through 2015. Whereas explanatory trials to establish whether a treatment works are performed under optimal situations with a narrow range of participants, Project ReD CHiP was a practical, real-life trial designed to improve hypertension detection and treatment, reduce hypertension health disparities, and overcome the barriers to care.[6] To accomplish these goals, the intervention employed several strategies, including training staff and patients in blood pressure measurement, management of patient care, and provider education. Six community-based medical practices that care for racially diverse patient populations within the Johns Hopkins Community Physicians network participated in the study.

At the beginning of the project, we conducted focus groups and interviews with patients, insurers, health care organizational leaders, providers, and frontline staff to better understand the needs of key stakeholders. We then developed a staged intervention with three components. The first featured improved blood pressure measurement training for clinicians and frontline staff along with the installation of new automated blood pressure monitors in clinics. The second gave clinicians training in patient-centered communication skills along with feedback from a user-friendly, web-based clinical dashboard about blood pressure control rates in their patient panels, broken down by race. The third component delivered care management for patients with uncontrolled hypertension by adding registered dietitians and pharmacists to their primary care teams.

The care management program included three sessions on-site at primary care practices, four weeks apart, for a total of two hours of face-to-face contact time. Three full-time registered dietitians at each site counseled patients to take their medications as prescribed and helped them set goals around health behavior change, including following the Dietary Approaches to Stop Hypertension or DASH diet (which features more fruits and vegetables, potassium, vitamin C, and magnesium—all proven to reduce blood pressure in trials of volunteers), losing weight, exercising, and adopting other self-management behaviors such as checking their blood pressure at home. A part-time pharmacist at each site helped to regulate the dosages of patients' medications between their provider visits and reinforced patients' adherence to medications and lifestyle change goals.

We provided the dietitians and pharmacists with training in cultural competence and motivational interviewing, which they used to assess and address patients' disparities-related barriers to self-management (for example, in areas of literacy, employment, finances, housing, or transportation). Care managers used telephone outreach to eligible patients identified through the electronic medical record, or providers referred eligible patients, identified during routine clinic visits, to the care managers. Because this was part of a research study, there was no cost to patients for participation.

To ensure that the study conditions were representative of real-life clinical settings, the dietitians and pharmacists who delivered the program were based at the practices, and providers continued their routine management of hypertensive patients; they collaborated with their care manager at their discretion.

Both African American and White patients who received care management showed dramatic reductions in blood pressure—down almost 10 points on average, and the program was equally effective among Blacks and Whites.[7] Patients who attended all three sessions showed the greatest improvement, but even those who attended only one or two sessions improved more than patients who did not participate in the program. In a separate analysis, we found that the ReD CHiP care management program was cost-effective from the perspective of the health care sector (for example, Medicare and private insurance companies) for preventing consequences of hypertension such as heart disease, stroke, or death among Black patients and patients 65 years or older.[8]

The program did have some limitations. First, care managers were unable to contact half of the eligible patients (this is a common indicator of social, financial, and health-related stressors in disparity populations, such as disconnected telephone services, evictions or displacements, hospitalizations, unexpected deaths, or incarcerations). Second, only 39 percent of the 629 patients who started the program completed all three sessions.

We concluded that the care management program succeeded in cost-effectively improving blood pressure control among patients who participated in care management, especially among African Americans and individuals over 65. However, because the program was based in clinics, not all those in need could benefit—for example, patients who had transportation problems or difficulty getting time off work might not have been able to attend visits. Thus, we suggested that African Americans might need greater support outside the clinic, and/or policies that address social determinants of health. We also suggested that program completion and reach might be improved if care management could be integrated within the daily routine of patients—for example, at faith organizations or employment sites—and that home visits or community health workers might improve reach and completion for high-risk and frail patients. These findings informed the design of the Center's subsequent studies.

The blood pressure measurement program, led by my mentee Romsai Boonyasai, was a great success. Clinics used the automated devices as intended in 72 percent of encounters, and doctors didn't repeat the readings as often as they did before—which indicated that they trusted the measures obtained by staff. Confidence in the measurements means the doctors are more likely to adjust patient medications as needed based on those readings rather than just falling back on a wait-and-see approach, leaving patients with no change in dosages they might critically need. The health system also continued to use the devices and our training program after the study ended.

Although the communication skills program was not used by most of the physicians in the study, the good news is that it is now being used in a health disparities skills program for internal medicine and family medicine residents and nurse practitioner students at the University of Washington. Its effectiveness will be tested in a clinical trial led by Janice Sabin, a clinical social worker and health equity researcher who worked with me on my earlier study on implicit racial bias, and Jennifer Tjia, a physician and epidemiologist, and their team will measure learner and patient outcomes pre- and post-intervention.

FIVE PLUS NUTS AND BEANS

Five Plus Nuts and Beans, a study led from 2010 to 2015 by my Hopkins colleagues Edgar (Pete) R. Miller III, a physician and epidemiologist, and Jessica Yeh, an epidemiologist, tested whether motivational counseling to follow the DASH diet, mentioned above, plus an income supplement to help with the purchase of healthy food from an online grocery store, or just the income supplement by itself, improved the health of participants.[9] Although pharmacological treatments of health issues such as hypertension that disproportionately impact African

Americans substantially lower their risk for disease, diet is also an integral part of managing health, although an often difficult one given the dearth of healthy food markets in African American and lower-income neighborhoods.

The Five Plus Nuts and Beans trial was an eight-week pilot study testing the effects of incorporating healthier diets among two groups of African Americans who were on stable doses of antihypertensive medications. We randomly assigned 123 African Americans with controlled hypertension from an urban primary clinic in Baltimore to one of two intervention groups and implemented the trial in partnership with a community supermarket and the Baltimore City Health Department. In one group, participants engaged in an initial one-hour session with a nutritionist who helped with the weekly purchase of potassium- and magnesium-rich fruits, vegetables, nuts, and beans from a community grocery store. These participants also received a weekly call from a dietary coach, weekly recipes and tip sheets about incorporating fruits, vegetables, nuts, and beans into their diet, and $30 of food per week for pick-up at their neighborhood library. The second group received nutritional advice and a $30-per-week food credit at a local supermarket, where participants made their own purchasing decisions.

One of the remarkable aspects of this study was the fact that it met 100 percent of its recruitment and follow-up goals. This is virtually unheard of in most research studies and especially in studies including groups that experience barriers to research participation, such as African Americans. We believe that this was due in large part to the excellent input we received from the Center's community advisory board about the messaging and appearance of the recruitment materials, the incorporation of the food allowance into both intervention groups, and the exemplary communication, cultural, and structural competence of the research interviewers and intervention staff. One example of this kind of competence—and dedication—comes from the Five Plus dietary coach, Debra Gayles. She tells the following story about one of the study participants who was assigned to the coaching arm of Five Plus:

> He was responsible for picking up his groceries at the Orleans Street Library. One day he didn't pick them up... he explained to me that he hadn't been taking his [psychiatric] medicines and was not able to participate in the weekly calls. I referred him to his doctor and asked him if he would mind if I ordered food for him since I had a record of which foods he liked. He agreed, and I continued to deliver his food to him. He thanked me for having patience with him while he was working out his mental health issues. . . . I was humbled that I could do a small part to show someone that I cared, as I was doing my job.

The interventions did not have an immediate effect on blood pressure, but they increased the consumption of fruits and vegetables and the urinary excretion of potassium, which showed that people were following the diet, and could positively impact their health over a longer period. This study also led to an ongoing follow-up study, called Five Plus Nuts and Beans for Kidneys, co-led by Dr. Miller and another Hopkins colleague, Deidra Crews, a kidney specialist with an interest in food as medicine who's also the Center's associate director for research development. The latter study, funded by the National Institute of Minority Health and Health Disparities, is looking at whether better diets can both lower blood pressure and protect the kidneys in people with early signs of kidney disease.

RICH LIFE

The third and more recent Center study is called Reducing Inequities in Care of Hypertension: Lifestyle Improvement for Everyone (the RICH LIFE Project).[10] This project, which runs from 2015 to 2021, is co-led by my Hopkins colleague Jill Marsteller, a health services and organizational behavior researcher, and me. Funded by a partnership between the Patient-Centered Outcomes Research Institute and the National Heart, Lung, and Blood Institute, it's designed to help lower blood pressure and heart disease risk among ethnic minority, low-income urban and rural populations. We've enrolled 1,822 patients and will continue to be in touch with them and collect information about their health and health care experiences for two years. The study is comparing blood pressure measurement and hypertension best practices training for staff, standard feedback on clinical performance of blood pressure control rates, and workshops for health system leaders with a much more comprehensive approach that includes a structured team approach to care and access to subspecialists or community health worker support as needed.

Thirty participating clinics including community health centers from across Maryland and Pennsylvania were divided into two groups. In both groups of clinics, clinical staff have received training in correctly measuring blood pressure; they get certified to do this on an annual basis. The staff members are also taking part in web-based training about taking care of people with high blood pressure. Finally, they're getting monthly feedback about how well their patients' blood pressure is being kept at a healthy level using what we call a hypertension dashboard; this information is provided overall and then separately for Black, White, and Latino patients.

In one group of clinics, in addition to the programs described above, primary care doctors, nurses, and staff members also receive more in-depth training that focuses on factors that cause people from socially disadvantaged groups to have a higher risk of uncontrolled blood pressure. This group is using a team approach to providing care. Primary care doctors, and other health professionals, such as pharmacists, social workers, and nutritionists if they are available at the clinic, all work together with a nurse care manager who delivers one-on-one education and counseling to patients to help them take care of their health. The nurse care managers get special training in motivational interviewing—a patient-centered communication approach similar to the one we used with physicians in my earlier studies.

With the second group, some patients may get home visits from community health workers who help them meet their needs related to transportation, financial or emotional stress, or housing or employment instability. The community health workers also get trained in motivational interviewing. In addition, some patients may have a group of specialists review their treatment plans and give recommendations to their primary care doctor to control their high blood pressure. These specialists include experts in heart and kidney diseases, diabetes, mental health, behavioral elements such as smoking, and high blood pressure. Regular case management meetings allow the team to discuss patient cases in depth and to brainstorm approaches to patient care. Meetings also allow the group to consider whether to bring in a community health worker to engage with difficult-to-reach patients or patients with complex social needs.

Educational programs for the health system leaders are also deployed to enhance the uptake and effectiveness of the other programs. This training includes webinars and telephone calls to disseminate evidence-based practices for health equity and facilitate dialogue among health system leaders and representatives from community organizations regarding efforts to advance health equity.

This study combines tested approaches with new ideas in an innovative program that treats patients as whole people, not simply treating a disease. It's looking at their quality of life and their ability to reach personal health goals from multiple perspectives. It's been designed to give doctors, insurers, and lawmakers the confidence to support similar programs in a wide variety of medical settings and among other at-risk populations.

The goal of the program is to help patients control their blood pressure, feel healthier, have more energy, and lower their risk of heart attack, stroke, or kidney failure, along with both recognition and better control of any coexisting conditions they might have. Our hope is that people will feel more in control of their lives, having set goals for themselves that they've been able to accomplish. They'll know how to monitor and track their high blood pressure. They'll understand the importance of letting their primary care team know of any concerns or needs that they have. They'll know more about the services that are available to them. I hope this program will help those who need it the most so that when it ends, everyone will be healthier, and those who were in the poorest health at the beginning will be just as healthy as those who were healthiest. Once the study is complete and we've shared our findings, we hope many other health care systems will adopt these strategies to address their own patients' needs.

In September 2020, we held an online town hall meeting about the RICH LIFE Project to which we invited patients, care managers, community health workers, and health system leaders to share an update about the project's progress, preliminary data, and lessons learned. One of our study participants, Mrs. Willi McNair, enthusiastically shared her experience:

> My sugar was bad, my pressure was high, but when I got with them [my RICH LIFE nurse care manager and community health worker] it was better. They gave me a cuff to take my blood pressure. That helped me. They gave me paperwork to keep my pressure down and my sugar down. They made sure that I had food and that I was okay about where I lived. They did come out to see me. They would call me in the evening to make sure I was on task; who does that? RICH LIFE does. When I started early on, I didn't listen, but then it hit me: These people actually care enough to come out, to call me, I better listen. My sugar came down, my pressure came down, [and] everything came into place. I would recommend them to anybody!

None of this work, and none of these success stories, would be possible without a highly trained, expertly coordinated team. My biostatistician colleagues, Kit Carson and Nae-Yuh Wang, have ensured that we used the best data collection and analysis methods for all of our studies, including this one. Kit spends countless hours checking the integrity of data, training our staff in how to properly document their work, and making sure we're successful in our recruitment and follow up efforts. The Center's also supported by talented and committed staff—notably Katie Dietz, our research program manager from the early years of the Center, Gideon Avornu who joined us for the RICH LIFE project, Jia Lee, the Center's administrative coordinator, and several dedicated research assistants. Our research staff's list of responsibilities is long: taking minutes during all meetings, calling study participants, writing checks and preparing bulk mailings, delivering equipment to clinical sites, supporting interventionists in their tasks, performing data entry and generating weekly reports, helping with blood pressure and other trainings, and organizing our study and community advisory board meetings. They're each vital gears of this machine.

LEADING RESEARCH WORTHY OF PUBLIC TRUST

Through the contributions of our staff, board, community partners, and participants, since 2015 the Center has moved beyond disparities in cardiovascular disease to examine other health disparities. I'd already been involved in research on depression, so it wasn't difficult to include depression among the conditions we study. Because disparities in many chronic conditions and their treatment share a common set of risk factors and contributors, including structural contributors to health inequities and resulting poor diets, obesity, tobacco smoking, sedentary lifestyle, and disparities in health care access and treatment, many of my colleagues who study disparities in diabetes, chronic kidney disease, cancer, and mental health conditions joined the Center. We were also joined by pediatricians and maternal health researchers who are helping us to broaden our work to examine intergenerational impacts of health disparities and to begin to take a life-course approach to our work.

We've begun conducting projects on a global scale rather than limiting ourselves to Baltimore and other areas of the United States. Our latest study, discussed in the next chapter, is being conducted in Kumasi, Ghana, not far from where I grew up in Liberia. There, we're adapting some of the lessons learned from our local studies in the United States to new contexts. We've seen that research is critical, but it can't solve problems in isolation; it has to be connected to people who can make changes in organizations and communities as a result of its findings. So, we've strengthened our focus on getting the results of our research out to broader audiences. We see our work as one of the ways we can help to repair the broken social contract between our society's institutions and communities of color and assure that the research we lead is worthy of public (and our local community's) trust. In 2020, the Center celebrated 10 years of working together to improve the health of the most disadvantaged in communities in Baltimore, the United States, and around the world, which positions us as nationally and globally influential scientists, practitioners, and leaders, as we'd hoped, and enables us to expand our work into influencing practice and policy.

CHAPTER 5

From Research to Practice and Policy

IN 2017, I WAS NAMED a Johns Hopkins University Bloomberg Distinguished Professor for my work as a health equity researcher and educator. Anyone working in academia would be proud of such a distinction, and I was (and am), but to me the new position represented something else—a tremendous opportunity to broaden the mission of the Johns Hopkins Center to Eliminate Cardiovascular Health Disparities, which the next year would become the Johns Hopkins Center for Healthy Equity. It raised the visibility of the Center's research across the university such that investigators in other Hopkins schools—such as the Krieger School of Arts & Sciences and the Applied Physics Laboratory, where I didn't previously have collaborations—knew they could reach out if they wanted to study something related to health equity. For instance, as we'll discuss in the next chapter, we partnered with Michael Degani from the Krieger School's Department of Anthropology, an expert in energy use and infrastructure in African cities, including the study of mobile apps, televisions and more, to address the reach of health care in Ghana. We're also collaborating with Brant Chee of the Applied Physics Laboratory on a grant application for research that would employ artificial intelligence to assess the speech patterns of physicians and patients recorded during medical visits—a means to develop software for use during future clinical evaluation programs—as a potential tool to address racial disparities in clinical communication. This new position allowed me to intensify my efforts to translate research into practice and policy to address the social determinants of health in vulnerable and socially disadvantaged groups, thereby moving beyond understanding the roots and mechanisms of these issues to achieving positive, systemic impacts on the health of communities.

In this chapter, I'll revisit some of the topics discussed earlier specifically from this perspective. How can we translate research into actions that make a difference in the lives of people everywhere, but especially those experiencing health disparities? Doing so requires that interventions be based on evidence and that researchers partner with communities, health systems, and other sectors of society to change practices. It also requires policies at local, state, and national levels, within the health care system and in other sectors, to explicitly reduce health disparities while also improving overall population health. In addition to equipping community members and organizations with skills to address these issues and to advocate for resources from payers and policymakers (as discussed in chapter 4), a significant part of this work is helping people in health care delivery systems and other sectors of society, including government, understand why they should care about health disparities—what their myriad and long-term impacts are that affect every one of us.

Practices to advance health equity can involve behavior change at the level of individuals, families and other social networks, organizations and institutions across various societal sectors, and even an entire community. In chapter 3, I discussed relationship-centered care and cultural and structural competence, two health care delivery practices to advance health equity. Here, I'll focus the discussion on practices related to health care and public health workforce development (through diversity and inclusion, training, including in bias reduction and anti-racism skills, mentoring, and leadership programs) and the use of health information and electronic communication technologies to advance health equity. Policies in nearly every sector of society could better support health equity. But leveraging policy to support health equity will require deeper understanding among policymakers of the conditions that entrench health inequities and the actions it would take to up-end this entrenchment at local, regional, national, and international levels. Successful policy changes will also require greater collaboration across sectors. Finally, addressing health equity through policy and advocacy requires proven approaches to improve health, evaluate social policies to identify health effects, and augment the workforce with people who have training in social and economic policy.

REFORMING THE PHYSICIAN WORKFORCE THROUGH DIVERSITY AND INCLUSION EFFORTS

By nature, people tend to be more comfortable with people who are like them. Our studies have shown that race concordance in the doctor-patient relationship (discussed in chapter 2) produces better communication and decision-making. Race concordance may also be associated with better access to care and better health outcomes as well (as reflected by a recent study documenting lower mortality among African American newborns receiving care from physicians of the same race).[1] Yet one of the most pervasive and enduring challenges facing the health care workforce is the critical shortage of racial and ethnic minorities serving in health professions. Diversifying the workforce may be one of the most important issues the health care system faces.

Historically, African Americans, Native Americans, and Latinos have faced severe barriers in gaining admission to schools of medicine, nursing, and dentistry and in securing careers in the health professions. Prior to the gains of the civil rights movement, Blacks were effectively barred from all but a few of the nation's medical schools and were systematically denied access to membership in state medical societies. Of the nearly 700,000 practicing physicians in 2012, only 9.2 percent were members of underrepresented minority groups.[2] Racial and ethnic minorities are represented in higher percentages among female physicians than among male physicians, but they remain vastly underrepresented. Today, Black women make up less than 3 percent of physicians in the United States.[3]

Beyond course-correcting the racial disparities, the numbers of underrepresented minorities in the health care workforce need to increase simply to meet the demand for health care professionals. Without action, the already dire shortage of health care professionals in the United States will continue to intensify, driven by an aging baby boomer population with more complex health needs, the geographic locations of patients relative to providers, and the increasing racial and ethnic diversity of the US population. By 2050—over the next generation—Whites will no longer be a majority population in the United States. Moreover, the number of older adults is expected to double by 2030—in 10 short years—at which point they'll constitute nearly 20 percent of the US population.[4] Racial and ethnic minorities, who already face severe disparities in health care delivery, are expected to make up an increasingly larger share of this population. New directions in workforce composition, distribution, expertise, and training will be essential to meet the needs of the changing population.

HEALTH CARE WORKFORCE DIVERSITY Percentages by race	AFRICAN AMERICAN	HISPANIC	WHITE	AMERICAN INDIAN or ALASKAN NATIVE	ASIAN	DATA SOURCE
US population	12.3	17.8	61.1	0.7	5.4	ACS DATA (2018)
Psychologists	4	5	86	<1	5	APA (2015)
Psychiatrists	7.3	6.0	77.9	0.58	13	AAMC (2018), tables 12 and 13
Primary care physicians	7.3	7.6	61.4	0.41	21.1	AAMC (2018), figure 26
Surgeons	6.3	7.5	48.0	0.48	17.3	AAMC (2018), tables 12 and 13
Nurse practitioners and Nurse midwives	8.7	5.5	77.5	0.52	6.2	American Community Survey, US Census Bureau (2014-19)
Registered nurses	6.2	5.3	80.8	0.4	7.5	2017 National Nursing Workforce Survey, table 4
Physician assistants	2.8	3.7	88.5	<1	3.9	Racial and Gender Disparities in the Physician Assistant Profession (2016)

Note: Row percentages may not total 100 due to missing data and persons classified in other race categories.

Those currently working in medical fields serve as role models for the next generations. Without physicians of color, we perpetuate a cycle of underrepresentation. As many have said, "You can't be what you can't see." So in addition to increasing the diversity of medical students, we need to increase the diversity of medical school faculty. Medical schools hold the mission of educating physicians who will care for the entire population. Diversity among faculty enhances the ability of academic medicine to fulfill its educational, research, and patient-care missions.

The Importance of Racial and Ethnic Diversity in Health Professions

People of African American, Latino, and Native American descent constitute 30 percent of the US population, but they are severely underrepresented in the US health workforce. As the US population becomes increasingly diverse, ensuring adequate workforce representation could:

- improve organizational excellence;
- increase access to health care for persons from underserved communities; and
- reduce disparities in health and health care.

Part of this professional underrepresentation stems from a similar imbalance in university and medical programs. Studies show that racial and ethnic diversity leads to improved academic and workplace environments, including:

- better educational experiences for medical students of all racial and ethnic backgrounds;
- higher ratings of preparedness to care for patients from other racial and ethnic backgrounds;
- more positive attitudes about equity and access to care;
- increased academic reputation;
- greater creativity and innovation;
- better problem-solving abilities.

African American, Latino, and Native American physicians are much more likely than their White peers to practice in underserved communities and to treat larger numbers of patients of color. While 100 percent matches of patients to health care workers who come from similar racial and/or social backgrounds isn't the goal (all professionals should be able to treat any patients regardless of their ethnicity), research has shown that patients of color in race-concordant relationships:

- report higher levels of satisfaction and participate more in decision-making;
- experience visits characterized by better communication and more patient-positive affect;
- report higher intentions to use preventive health care services
- have higher levels of adherence to medications for heart conditions;
- are more likely to be prescribed HIV medications in a timely manner; and
- have significantly better health outcomes in some studies.

FURTHER READING

W.D. King, M.D. Wong, M.F. Shapiro, B.E. Landon, and W.E. Cunningham. 2004. Does racial concordance between HIV-positive patients and their physicians affect the time to receipt of protease inhibitors? *Journal of General Internal Medicine* 19(11):1146–53.

A.H. Traylor, J.A. Schmittdiel, C.S. Uratsu, C.M. Mangione, and U. Subramanian. 2010. Adherence to cardiovascular disease medications: does patient-provider race/ethnicity and language concordance matter? *Journal of General Internal Medicine* 25(11):1172–7.

Marcella Alsan, Owen Garrick, and Grant C. Graziani. 2018. Does diversity matter for health? Experimental evidence from Oakland. National Bureau of Economic Research Working Paper series no. 2478, https://www.nber.org/system/files/working papers/w24787/w24787.pdf.

M.J. Shen, E.B. Peterson, R. Costas-Muñiz, M.H. Hernandez, S.T. Jewell, K. Matsoukas, and C.L. Bylund. 2018. The effects of race and racial concordance on patient-physician communication: A systematic review of the literature. *Journal of Racial and Ethnic Health Disparities* 5(1):117–140.

B.N. Greenwood, R.R. Hardeman, L. Huang, and A. Sojourner. 2020. Physician-patient racial concordance and disparities in birthing mortality for newborns. *Proceedings of the National Academy of Sciences* 117(35): 21194–21200.

Having underrepresented minority faculty members in medical schools promotes more effective health care delivery to a diverse population, improves the quality of medical education for all races and ethnic backgrounds in attendance, and stimulates research attentive to the needs and concerns of minority groups. Studies have shown that more diverse faculties also bolster levels of innovation and problem-solving, and enhance the academic reputation of their institutions. Diversity is not simply an ethical issue but an excellence issue.

Despite these compelling reasons for diversifying health care faculty, there's an alarming lack of minority faculty in US medical schools and an especially serious dearth in leadership or senior roles. Studies have shown that underrepresented minority faculty are more likely to advance more slowly and to report the intention to leave academic medicine.[5] In addition, minority faculty members report experiences of ethnic harassment, biased treatment, and racial fatigue.[6] They spend more time focused on patient care and less on research than their nonminority colleagues. Sustained efforts have increased the enrollment of underrepresented minority medical students, but the environment for underrepresented minority faculty has received much less attention and is equally vital in increasing the diversity of the health care workforce.

Having more underrepresented minority faculty in senior and leadership roles in medical schools could increase the cultural awareness and skills of all physicians-in-training and biomedical scientists.[7] The result would be a greater capacity to care for underserved groups and to determine the causes of and solutions to health disparities. Fully engaging the skills and insights of these faculty members, beyond the ethics of inclusion, achieves the best possible science and medical care.

There are many ongoing efforts to improve structural factors (the diversity of the faculty, trainees, staff and leaders, and the mission, policies, resources, and practices of institutions) as well as psychological and behavioral factors (social interactions among individuals and groups and perceptions of bias, discrimination, and inclusion) within academic medicine. One example is a series of workshops called "Breaking the Bias Habit," based on research by Molly Carnes and colleagues at the University of Wisconsin–Madison, that introduces academic audiences to the concepts of implicit bias (discussed in chapter 2 and later in this chapter) and stereotyping about diverse groups of people using many of the same practical steps I described earlier.[8] However, in this case, the focus is having interactive discussions about the potential influences of implicit bias in their own academic units and applying evidence-based strategies for reducing the application of these biases within academic practices (for example, teaching and evaluating learners, interpersonal micro-aggressions, recruitment and promotion of women and people of color). But until such worthy and effective efforts become core parts of missions, such initiatives, hampered by insufficient leadership support and resources, remain siloed. And, without collective responsibility to drive the change, ultimately they fail to be implemented and sustained.

Another effective effort is a series of programmatic changes in the medical school admissions process undertaken by Quinn Capers and his colleagues at Ohio State University.[9] These changes include having an explicit diversity mission, using anonymous voting, expanded screening and a holistic review of applicants, blinding of reviewers to academic metrics, provision of implicit bias training to reviewers, removing applicant photos during committee discussions, and diversifying committee membership. Other programs are focused on building the pipeline of potential applicants from underrepresented communities starting in secondary school through college and beyond, although these programs continue to face challenges associated with anti-affirmative action legislation.[10]

TRAINING THE HEALTH CARE AND PUBLIC HEALTH PRACTITIONER WORKFORCE

Increasing the racial and ethnic diversity of health professionals has great potential to improve patient-physician communication and patient outcomes to address disparities among vulnerable populations. However, not every patient will be able to have a physician who is their same race. For that reason, the training of health care personnel needs to prioritize cultural and structural competence as strategies to advance health equity, as discussed extensively in chapter 3.

Many health professional and public health schools have struggled with integrating health equity more effectively into their curricula. Though many schools now offer introductory or elective courses on health equity, this topic needs to be woven throughout all aspects of health professional and public health education. Regardless of what topic is being taught, whether nutrition, heart disease, kidney disease, or mental health, these schools need to address health equity issues and how they affect people of different backgrounds to fully prepare their students for all patients they'll encounter. The social circumstances that contribute to these problems need to be discussed, as does the role that these professionals can play in reducing and eliminating these problems. The schools and their associated continuing education programs should incorporate competency-based curricula that include practices for working in interprofessional, multidisciplinary teams, interpersonal and organizational approaches to addressing health equity, and partnership-building skills for working with community-based organizations and across societal sectors.

TRAINING THE HEALTH EQUITY RESEARCH WORKFORCE

Research has repeatedly revealed that the contributors to health disparities are complex and that they interact in different ways between and within groups, settings, and contexts. This complexity calls for the practice of training a cadre of researchers who represent multiple disciplines and multiple sectors of society.

Under the leadership of one of my mentees, Tanjala Purnell, an epidemiologist and health services researcher whom I appointed as associate director for education and training at the Center, we've developed a mentoring and training curriculum that we hope will serve as a model for other centers and institutions across the country and around the world.

Early on, our prospective trainees told us that they needed practical tips about what it was like to conduct health equity research in the field, so we created curricula for the Bloomberg School of Public Health and the School of Medicine at Hopkins that involve lectures with lessons learned from experts in the field as well as real-world experiences in research team meetings and in clinical and community settings whenever possible. We also learned that many high school, undergraduate, and graduate students from disciplines outside of medicine, nursing, and public health, and from institutions across the country and the world, wanted access to these experiences, so we created internships and electives, degree-level courses, and online programs to enhance our reach.

We've developed a process for connecting potential trainees with faculty mentors who are willing to contribute their time. To do this, we created a database of faculty interests, projects, and opportunities for trainees that we update quarterly. When trainees express interest, we can link them with the appropriate faculty member. Once trainees begin to work with their mentors' program, they each complete a mentored research project. We also hold "meet the professor" sessions and career development panels where graduate students at Hopkins share practical tips about what they did to get accepted to the university and how they've been successful. We include field experiences with community-based organizations, partly by leveraging existing partnerships with members of our community advisory board, and all trainees are invited to attend our community advisory board meetings as well as the research team meetings of their mentors.

Since 2017, Dr. Purnell and I have developed two courses for academic credit called "Applications of Innovative Methods and Health Equity Research" and "Local and Global Best Practices in Health Equity Research." These courses include lectures, case-based examples, interactive panels, and discussions. They feature faculty in the Center for Health Equity and across Johns Hopkins who have substantial expertise in the subject, as well as health equity experts from other institutions in the United States and around the world. Our first course has earned the Excellence in Teaching Award several years in a row, and we hope we will have the same positive results for the second course, which was launched in 2020.

We've also created two massive open online courses (MOOCs) on the Coursera platform. "Foundations of Health

Equity Research" introduces students to the core principles of health equity research. "Application of Health Equity Research Methods for Practice and Policy" is intended for students who have completed the foundations course or have previous experience in health equity research, practice, or policy.

Many training programs focus on increasing knowledge; fewer focus on developing skills and changing attitudes among learners. Yet these attitudes and skills are particularly important in doing research to address health disparities. Before we developed our courses, we heard many research trainees complain that the content of their courses that covered health disparities had no practical applications for their day-to-day work. Some trainees also stated that their educational environment wasn't welcoming, which is particularly important for people who come from disadvantaged backgrounds. Training needs to occur in a safe space where learners can explore their attitudes and beliefs and where they can acknowledge their need for additional knowledge or skills without being judged or shamed. Presentations need to be engaging and discussions respectful of the life experiences and expertise of the learners. Training programs should allow trainees to learn from one another. We developed our program with all these factors in mind.

The Center also launched a monthly "Health Equity Jam Session" seminar series in 2015 to provide an informal opportunity for colleagues, trainees, and community partners to collaborate and share their work. With an emphasis on developing future generations of health equity scholars, the jam sessions provide a supportive forum to discuss research ideas, proposals, research in progress, responses to peer review, career development, and collaborations. Many research collaborations have come out of these sessions, which are also used for career development and professional networking. For instance, we once held a session using a speed-mentoring approach that allowed trainees to chat informally with faculty members because we learned of their mutual interest in meeting. We've now instituted a special annual session on health equity training opportunities at Johns Hopkins in partnership with the Urban Health Institute and other centers across the university.

Jam sessions have included speakers on topics such as toxic stress as a social determinant of health, behavioral health treatment for homeless persons, addressing food insecurity, building community-academic partnerships, best practices for engaging policymakers, disparities in maternal health, and local and global learning in health equity research and practice. In 2019, we co-hosted a special symposium with the Johns Hopkins Bloomberg School of Public Health, the Department of the History of Medicine in the School of Medicine, and the Urban Health Institute, called "1619-2019: The Legacy of Slavery for Health Equity in Baltimore and Beyond." All the sessions are livestreamed to allow viewing from any location, and we post an archive of the sessions on the Center's YouTube playlist as well as our Periscope Account to allow 24/7 access. The idea is to create larger and larger communities of people with diverse interests, skills, empathy, and expertise to have the greatest possible impact on health disparities.

We believe that educational programs targeting health equity should train practitioners, researchers, and lay community members—especially those from disadvantaged communities—alongside each other to enhance their skills in intervention design, delivery, and evaluation, and in leadership and advocacy skills, to bring about social change. We also believe program evaluations should use methods that incorporate learner behavioral outcomes and population health outcomes and that examine program influences on reducing health disparities, and that's what we are doing with our program.

Our unique trainee mentoring program has helped to launch more than 150 scholars, including high school and undergraduate students, medical, nursing, and public health students, postdoctoral fellows, and junior faculty. Our degree-level courses at Johns Hopkins have educated more than 120 students so far, and our open online courses have trained more than 2,300 people across the globe in 2020 alone. Our jam sessions have attracted 7,500 attendees over the past five years.

BIAS REDUCTION AND ANTI-RACISM TRAINING

Educators have proposed a framework for integrating implicit bias recognition and management into health professions education that's based on previous research and includes the following practical actions for curriculum developers: (1) creating safe and nonthreatening learning settings, (2) increasing knowledge about the science of implicit bias, (3) emphasizing how implicit bias influences behaviors and patient outcomes, (4) increasing self-awareness of existing implicit biases, (5) improving conscious efforts to overcome implicit bias, and (6) enhancing awareness of how implicit bias influences others.[11] This approach can be used for practitioners and researchers.

At Johns Hopkins, we provide training to medical students on implicit bias early in their first semester and to students in our masters of health administration program, using the approaches mentioned above as well as some described earlier, in chapter 2. We've also expanded the medical student curriculum to include what's been called bystander training in the past but what we now call upstander training. We know that the tone of an environment drives behavior. Therefore, we've suggested that upstander anti-racism training, which has been used by laypeople to respond to incidents of interpersonal or systemic racism, has potential as an organizational strategy in academic medicine and health care, public health practice, and research.[12]

In previous research, actions resulting from this type of training have led to positive effects for the recipients of derogatory behavior, the "upstanders," as well as the persons who are responsible for the derogatory behavior.[13] If people are being treated differentially or disrespectfully because of their race, gender, or any other characteristic, other people in the environment have to be trained to respond or to "stand up" and speak out in defense of the person or people on the receiving end of derogatory treatment. If the prevailing culture is to not confront this type of behavior or to say anything, the norm becomes that it's acceptable to behave that way. Actions addressed by this training include confronting the "actor," recruiting other active upstanders, supporting persons from targeted groups after an experience of discrimination, formally reporting the incident, and/or seeking assistance from persons with authority to take action.

Implicit attitudes are more likely to influence behaviors when the capacity for cognitive processing is low. This capacity can decline for a variety of reasons, including fatigue, pressure, and cognitive overload. Policies and practices designed to mitigate stress and burnout, can simultaneously improve practitioner or researcher mental health and cognitive processing capacity while also enhancing service delivery, research, and patient or research participant outcomes.

One way to reduce stress and burnout on an organizational level is for leaders to implement policies and practices that create a culture of equity, civility, and respect, regardless of the roles and social identities of individuals within their institutions. Another promising way to do this on an individual level is mindfulness practice. Mindfulness can reduce the likelihood that implicit biases will be activated in the minds of practitioners and researchers, increase their awareness of their own implicit biases, enhance their ability to manage their responses to these biases once activated, increase their compassion toward themselves and empathy toward their patients, research participants, and community partners, and reduce internal sources of cognitive load such as stress and burnout.[14]

MENTORING

Mentoring is a vital component in training the next generation of health care and public health professionals and health equity researchers; it's a form of the ancient master and apprentice relationship. Excellent mentors have several key qualities: they socialize and nurture their protégés, teach them professional norms and expectations, impart specific knowledge, and give constructive feedback. Mentors motivate and inspire others and serve as role models: they show by what they do. They also advocate for their protégés. One of their most important roles is to open doors for their mentees and help them navigate complex institutional and personal issues. My colleague Spero Manson, Distinguished Professor of Public Health and Psychiatry and director of the Centers for American Indian and Alaska Native Health at the University of Colorado, shared the following story with me that exemplifies the value of having diverse role models in health-related research to inspire the next generation. Dr. Manson said,

> I am one of 67 grandchildren, the eldest male, born to a large extended family from the Turtle Mountain Chippewa Reservation in North Dakota, and the only one from my generation to complete a post-secondary education. . . . During family gatherings, though, a favorite aunt would make comments about my seemingly delinquent study efforts over a 23-year education. But my grandmother would fly to my defense: "Leave him alone! I'm a student, too, a student of life. We must never stop learning." Like my grandmother, I have come to champion life-long learning to enhance social inclusion, to promote active citizenship, and to stimulate intellectual development. I have been privileged to mentor more than 150 younger colleagues, Native and non-Native alike. My work with these nascent scientists emphasizes the importance of persistence throughout their journey, which, in the words of my grandmother, is most fulfilling when life-long.

I know how important it is to be a good mentor, because mentors helped lead me to where I am today. When I was trying to decide whether to be a clinician, an educator, a researcher, or all of the above, I sought out many people on different career paths who were doing things that I might like to do in the future. It wasn't always easy. Sometimes you meet people with whom you can connect but they don't know anything about the topic that you're trying to study. I tried to align myself with people who could be helpful in terms of both the big picture and their more specific expertise. From my parents to the other mentors I've mentioned, I would not be who I am or where I am today without their guidance and examples.

Mentors need to set expectations early in the relationship. Also, you can't mentor too many people simultaneously—you need to be realistic about how much you can take on so that you're truly available to your mentees. One way I've done this is by grouping my mentoring activities into certain times of the week. This gives me the time to focus on my mentee's issues and not have to do it while I'm in the midst of other things.

Just as my mentees learn from me, I've learned from them. Each has taught me how to be a better listener and a better advocate. I've learned about the value of caring for them not just as scientists, researchers, and future physicians but also as friends and colleagues. I've learned to respect and value their unique perspectives and contributions even if we're not always in agreement. I'm grateful for the honor of being able to work with every one of them. Just as I encourage them, I often tell them that some days they're the reason I get out of bed and continue to do this work. Their energy, enthusiasm, and optimism keep me hopeful about the future.

Through mentoring, I've learned that when I help others achieve their dreams, we all enjoy the journey. No single person can solve the problem of health disparities—the problem will need to be addressed for generations to come. As a woman and person of color, I feel a keen sense of indebtedness and responsibility to honor the legacies of my own mentors and to pay forward the knowledge, wisdom, and support I've received. I know that mentors serve as important role models for the next generation in a field where we desperately need the talent of people of color, and I'm honored to have mentored more than 75 individuals since joining the Johns Hopkins faculty, many of whom are women and people of color. Most of my former mentees who have completed their training have academic appointments and work as researchers or educators; others are health system clinicians, administrators, public health practitioners, and social entrepreneurs. They've won national awards and grants, and several have leadership roles within their organizations, as division directors, deans of schools of medicine and public health, hospital administrators, and public health commissioners.

LEADERSHIP TRAINING

Another lesson that I've learned about implementing practices and policies, informed by research on what is most likely to be effective, relates to the need for leadership. Leaders create change, mobilize stakeholders, and advance solutions to issues of significant concern. They can improve coordination, collaboration, and opportunities for soliciting community input on funding priorities and involvement in research and services. Leaders in government, in educational institutions, and in health care organizations need to help change narratives about health disparities and implement successful programs and policies to reduce or eliminate them.

There's a constant demand for new leaders with fresh viewpoints and energy who will fight for health equity. In particular, the growing pool of talented leaders among racial and ethnic minorities and in underserved communities could share their lived experiences and use their local wisdom to raise awareness and take action against health disparities. Future generations can be groomed for leadership by engaging and including them in the planning and execution of health, wellness, and safety initiatives.

One example of such a program is the Bunting Leadership Program, now under the umbrella of the Johns Hopkins Urban Health Institute, which I was appointed to lead in April 2020. Established in 2016, this year-long program enhances the capacity of young, passionate Baltimore community advocates with the skills to help improve the trajectory of health in their communities. The program includes experienced community leaders, academics, and business owners as faculty and incorporates peer support, reflective learning, and faculty guidance. The curriculum also incorporates content on the history of Baltimore, theories on leadership and community development, research and policies, and evidence-based practice.

USING TECHNOLOGY TO CHANGE INDIVIDUAL, SOCIAL NETWORK, AND ORGANIZATIONAL PRACTICES

Novel technologies have enormous potential to promote health equity, both in developed countries like the United States and around the world. Such innovations include "telemedicine" or video doctor visits, text reminders to take medicine or exercise, fitness tracking devices, heart rate monitors, disease monitoring devices, and virtual health advocacy platforms. New technologies run the risk of widening health care disparities by working primarily for communities that already have digital resources. But they also have the potential to reach and engage communities that typically have been underserved by the health care system, especially in areas with low percentages of health care workers.

Many of these technologies can be accessed through a smartphone, and ownership of smartphones has grown across the economic spectrum. African Americans, for example, have smartphone ownership that's nearly identical to that of the general population (80 percent versus 81 percent, respectively).[15] Moreover, racial and ethnic minorities in the United States use more smartphone apps than do Whites and are more likely to use their smartphones to access health information.

Unfortunately, relatively few culturally informed or culturally useful health informatics or digital health interventions or tools are available today. Innovations that fail to account for the needs of diverse users may largely benefit health outcomes in one group or sector of society while inadvertently creating, perpetuating, or increasing health disparities in another, thereby further reinforcing health inequities.

Fostering African-American Improvement in Total Health (FAITH!) is one example of a successful mobile intervention that facilitates access to health promotion resources.[16] Led by LaPrincess Brewer, a cardiologist at Mayo Clinic and one of my former Hopkins mentees, the FAITH! Study was designed to improve the cardiovascular health of African Americans. This intervention took the form of an in-person, church-based health education program. Three African American churches in Rochester, Minnesota, participated in this culturally tailored 10-week education series that incorporated the American Heart Association's Life's Simple 7 framework, an evidence-based metric of seven health-promoting behaviors and biological factors that improve health outcomes. Together, researchers and community members created a culturally aligned intervention that had a positive impact on the cardiovascular health of the study participants. On the basis of this success, the academic-community partnership secured federal funding to expand the reach of the FAITH! intervention by creating a mobile app designed to maximize acceptability, usability, and satisfaction among African American users. Dr. Brewer says she chose to partner with congregations to develop and test this program because "the black church is the premier institutional backbone of the African American community. . . . These churches have been a historical source of aid, salvation, and health services for underserved populations."

By tapping into existing social and community networks, FAITH! supports marginalized populations through collective mobilization and enhancement of resources, reduction of social isolation, and sharing of knowledge. It represents a technology-mediated solution to promote positive health behaviors.

New technologies, particularly smartphone apps that can be continually improved and updated, offer the advantage of reducing the lag between research and application. When new solutions to improving health for disadvantaged populations are discovered, technologies can be both widely disseminated and individually adapted. In this way, smartphone technologies could have immense impacts on populations around the world.

New technologies can also take advantage of built-in data collection to boost understanding of the mechanisms driving health disparities. By revealing key differences between groups, information systems could allow future interventions to be designed with these differences in mind. In the face of increasing reliance on telemedicine since the onset of the COVID-19 pandemic, experts have called for a multifaceted approach to policy, design, and implementation of these health technologies.[17] They say this approach must include assessment of patient access to technology as a standard of health care, the inclusion of sociodemographic and literacy measurement standards, involvement of diverse potential users in design phases, and the application of federal equity mandates, such as the National Standards for Culturally and Linguistically Appropriate Services (CLAS), to digital health.

Our country has watched as social media platforms have helped enact political and social change. As we've seen with the Dreamers and Black Lives Matter movements, these platforms have been particularly helpful in mobilizing young people who are passionate about issues of social justice, allowing them to reach audiences that range from leaders of academic institutions to corporations to government. Compared to White persons, Black and Latino persons have equal or greater rates of social media platform usage and are more likely to report using social media sites to get involved with issues that are important to them. Information technology can also be a fast, efficient, and far-reaching way to inform people about the latest research, including newly identified diagnostic tests, treatments, and vaccines. Communities and health care systems can learn what others across the nation and around the world are doing. The right media can reach groups of people who are often harder to reach, including racial and ethnic minority communities, young people, older persons, persons with disabilities, LGBTQ groups, and geographically isolated individuals. However, the sources of information and the nature and accuracy of messaging are critically important in preventing misinformation, which can lead to harmful social reactions and perpetuate existing mistrust of science and health care.

ORGANIZATIONAL AND GOVERNMENTAL POLICIES TO PROMOTE EQUITY

All the steps I've discussed in this chapter—and many more—require enabling policies. Policies that drive broad changes at the level of states or the entire country include laws and regulations in sectors such as housing, employment, education, transportation, food and nutrition, the environment, and health care—that's policy with a large *P*. One example of a set of policies with the potential to advance health equity includes laws determining federal and state minimum wages, which may have disproportionate effects on women, African Americans, and Hispanics who hold low-wage, but often essential worker, jobs. Evidence to support raising the minimum wage comes from studies suggesting that these wage increases reduce rates of smoking among pregnant women, low birthweight among infants, and absences from work due to illness, among employed adults.[18] Another example comes from laws to regulate the density of liquor stores, bars, and restaurants serving alcohol, such as *TransForm Baltimore*, a major rewrite of the Baltimore City zoning policy. Evidence to support this law comes from research showing that alcohol outlet density within a geographic area or per population can contribute to increased violence, crime, traffic accidents, and injuries.[19] Because low-income communities and communities of color have been found to have higher concentrations of alcohol outlets than wealthier areas, this law has the potential to reduce health disparities. My mentee, Rachel Thornton, a former White House Fellow, led the health impact assessment of the zoning rewrite.[20]

The 2010 Affordable Care Act (ACA) provides far broader access to health care and protects people against unjust refusal of coverage caused by preexisting health conditions. After the Act took effect, 3 million more African Americans, 4 million more Latinos, and nearly 9 million more White adults became insured. These gains lowered the US uninsured rate from 16 percent to 8.6 percent. Between 2013 and 2016, all racial and ethnic groups experienced gains in health coverage, especially minority groups and individuals with incomes below 139 percent of the federal poverty level. A recent review found significant evidence that the ACA has reduced social disparities in health care access, and expanded the use of primary care, and that it had a significant impact on the volume and range of services offered and the financial security of community health centers that accompanies the increase in insured patients.[21] Where evidence for the ACA's impact on health outcomes exists, it suggests that Medicaid expansion is the part of the ACA that has had the greatest impact on health disparities including preterm births and mortality. This suggests that a public option, which lowers the out-of-pocket costs for individuals, may also be a successful strategy to reduce health disparities.

Expansion of health insurance coverage has the potential to benefit not just individuals and groups but our entire society. For example, emergency room visits from the uninsured drive overall health care costs, and improving access to preventive and primary care services reduces these costs, which reduces the overall financial burden to taxpayers. Both modest policies and more complex, far-reaching ones have roles to play in changing our health care system or in changing other practices and behaviors that impact health.

We traditionally think of policy as something imposed by government, largely through legislation and rulemaking, seemingly behind closed doors. But institutional policies within the private sector, which I call policy with a small *p*, can also have a powerful influence on health. For example, a health care system might require its staff and providers to collect and examine data on quality metrics such as blood pressure control, control of diabetes, and COVID-19 infection rates among those tested, stratified by race and ethnicity of the patients served, and it might provide recognition or financial incentives to providers who reduce disparities in the care of their patient panels.

At the Center for Health Equity, we're committed to bringing research into policy decisions within private institutions, at all levels of government, and across sectors. We're working to inform and support the enactment of policies that reduce or eliminate unhealthy neighborhood or workplace conditions, support healthy behaviors, make the healthy choice the default choice, and promote equitable access to high-quality diagnostic and treatment services. We participated in an expert advisory group to highlight the impact of disparities in Maryland and offer evidence-based actions to help combat their effects. This led to the passage of the Maryland Health Improvement and Disparities Reduction Act of 2012, which enacted Health Enterprise Zones—local coalitions that qualify for special tax credits to attract providers to low-income communities and other benefits to implement health promotion programs and improve health outcomes. The Center's most recent policy initiative, led by Dr. Thornton, who is also our associate director for policy translation, is a partnership with the Bloomberg American Health Initiative at Johns Hopkins and IBM Watson Health to develop a method to measure the impact of hospitals on community health and equity, for potential inclusion in the Fortune/IBM Watson Health 100 Top Hospitals Program. Adding a measure of how hospitals contribute to community health and equity alongside measures of health care quality and patient satisfaction in the Fortune/IBM Watson national hospital ranking—and counting it equally in overall rankings—could recognize and reward a growing number of hospitals that are committed to investing in and improving population health and health equity in their own communities. It could also provide an incentive for other hospitals to change their policies and practices. The proposal, which has four components—population-level outcomes, hospital as health care provider, hospital as community partner, and hospital as anchor institution—was posted for public comment in August 2020, and we're looking forward to the next steps in this effort.[22]

Thinking about how to change policy, especially on a national level, can create a sense of frustration and hopelessness. However, solutions don't depend exclusively on action at these larger scales. There's a balance to be struck between the roles of local, state, and federal officials. As discussed in chapter 4, initiatives to reach underserved populations often need to be based on guidance and input from members of these communities, from community-based organizations, and from leaders who have knowledge of the needs and available resources within vulnerable communities. To be effective, these local initiatives generally require funding and encouragement from state and federal leaders. The success of all of these solutions can inform broader change, even on an international scale.

CHAPTER 6

A Global Perspective on Health Equity

HEALTH INEQUITIES OCCUR ON ALL SCALES, from the local to the global. Because of my early life experiences growing up in Liberia, attending school in Switzerland, and then coming to the United States, I've always had a global perspective on both health disparities and the role of relationships across social, cultural, and ethnic differences in addressing those inequities. After receiving the MacArthur Fellowship, I began to travel and learn more about how countries around the world approach health inequities. I discovered that in spite of the universal health care system in the United Kingdom, disparities in health and health care exist for Black, Asian, and minority ethnic groups relative to Whites in that high-income country. As a member of the People to People Citizens Ambassador Program Quality of Health Care Delegation to South Africa (a middle-income country) in 2009, I saw firsthand the inequities in health and health care there among Blacks, Indians, and Colored people, and Whites. Interestingly, on a trip to Cuba with the Medical Education Cooperation with Cuba Program in 2010, I learned about how a low-income country with a strong focus on prevention and primary care could be successful at minimizing health inequities. But my most significant international opportunities were yet to come.

The West African Ebola virus outbreak from 2013 to 2016 took a drastic toll on human life. It also resulted in the collapse of already fragile health systems, impacting economic growth and food security. Working across sectors and continents, several government and nongovernmental agencies mobilized to respond to and contain the epidemic, trying to help those afflicted, and to keep it from crossing their thresholds. Of particular relevance to this terrifying crisis was the need for health professionals to volunteer their time and skills to care for the victims of a highly contagious illness. They would share their expertise with other frontline workers, co-lead public health campaigns to try to contain the spread, and advocate with policymakers to address not only the current crisis, but also the long-standing inequities in access to care and quality of care in the countries that bore the largest burdens of the epidemic.

The outbreak inspired me to advocate for health equity with global partners. I began by obtaining funding and personal protective equipment for Last Mile Health, a global organization founded by my colleague Raj Panjabi, on which I serve as a board member. Last Mile supports frontline health workers and leaders to strengthen rural community health systems around the world. Dr. Panjabi and Last Mile (recipients of the $1 million

TED Prize in 2017) partnered with the government of Liberia and other organizations to design and scale a National Community Health Assistant Program which has trained and equipped thousands of community health workers, nurses, physician assistants, and midwives serving the majority of the rural population in Liberia.

Then in early 2015, I agreed to serve as chair of the data safety and monitoring board (DSMB) for PREVAC—the Partnership for Research on Ebola Vaccinations—which continues to oversee studies in Liberia, Sierra Leone, Guinea, and Mali. A data safety monitoring board is made up of outside experts who monitor research participant safety and the efficacy of the study product while a clinical study is taking place. They do this by providing an ongoing independent review of data from the trial. I also chaired the DSMB for PALM: a multicenter, multi-outbreak randomized, controlled safety and efficacy study of investigational therapeutics for the treatment of patients with Ebola virus disease during the world's second-largest Ebola outbreak on record, which occurred in active conflict zones in the Democratic Republic of Congo from 2018 to 2020. Global scientific partnerships that are established to test new vaccines include academia, pharma and biotech industries, communities, and policymakers such as the World Health Organization. This work not only connected me with several other global health researchers but also showed me that my expertise in epidemiology (the study of distributions and determinants of health-related states and events in defined populations), health disparities, community-based participatory research, and clinical trials in under-resourced settings in the United States was scientifically relevant, and that my life experiences growing up in West Africa made me particularly well suited to serve as a trustworthy and culturally knowledgeable leader. I was also grateful for the opportunity to serve my home country (and other African countries) at such an important time.

However, it was not until the Center for Health Equity had completed its first five years and I was named a Bloomberg Distinguished Professor in 2017 that I finally had the resources and networked collaborations to expand my own research and educational efforts to include global health settings. I hired a program director for operations and strategic initiatives, Nancy Edwards Molello. At that point, the Center began to leverage its experiences in team science and building strong community-academic partnerships toward a bold new vision of global health equity. We call this our local-to-global and global-to-local initiative, which has already started to contribute knowledge about how to use mobile health technology and community-based nurses in low-resource settings in West Africa to enhance access to care for patients with chronic medical conditions, and best practices in building community academic partnerships to advance health equity.

Researchers and public health officials from resource-rich nations often feel they have all the answers. But if they fail to listen to community leaders, they can't learn what interventions could have the biggest impact. Mrs. Molello recalls that on a trip to West Africa in 2014, she toured a tent designed for treating Ebola, installed by a well-intentioned nation, at a cost of a million dollars. It now sits empty and unused because the community can't afford its upkeep and because its design didn't resolve their need for a traditional health care facility. "The whole idea of both groups working together didn't happen," she said. "I think that's what global-to-local really stands for: global and the local communities working together to solve problems to help the world."

HYPERTENSION IN AFRICA

The Center's Addressing Hypertension Care in Africa (ADHINCRA) study is a prime example of the bidirectional local/global applications approach we've been taking. The study focuses on techniques to manage high blood pressure, which is a major cause of mortality in western Ghana, a region with a severe shortage of physicians and nurses. When creating the hypertension treatment protocols, the ADHINCRA team adapted those of the World Health Organization for treating high blood pressure to make them more suitable for the Ghanaian context. Ghana had just nine nurses and one physician per 10,000 people in 2015, compared with 49 nurses and 19 physicians per 10,000 people in the United States. The goal of the study, which is ongoing, is to explore whether blending technology, community-based caregivers, and culturally relevant messaging can improve outcomes in the management of high blood pressure for low-income patients. The study's name and logo were inspired by the Ghanaian Adinkra symbol that means "Akoma Ntoaso" (linked hearts) and is a symbol of unity and agreement. Because this project is a collaboration between Johns Hopkins University and Kwame Nkrumah University of Science and Technology (KNUST) in Ghana, we thought that it was an appropriate logo for the project.

Leveraging a local/global learning approach, the ADHINCRA study combines new ideas with approaches tested in the United States, in some of the Center's earlier studies, in an innovative program that treats patients as whole people. The study combines culturally appropriate patient management and messaging with modern blood pressure monitoring technology to help people manage chronic diseases. The intervention is especially vital in light of the World Health Organization's estimate that up to 75 percent of deaths in sub-Saharan Africa will be attributable to hypertension by 2030.[1] Developing scalable interventions in this region could improve the well-being of millions of people.

The study has sought to leverage the ubiquity of smartphones among the Ghanaian population. We partnered with Medtronic, a medical device company, to tailor one of the mobile phone apps they developed, called Akoma Pa, for our project. This platform is designed to enhance clinical decision support, shared decision-making, participatory communication, knowledge, treatment adherence, and self-monitoring of hypertension. The Bluetooth-enabled app delivers motivational messages designed specifically for West African notions of the body, responsibility, and well-being. Examples of healthy motivational messages would be to dance in order to be physically active, to cook vegetables without adding salted fish and meat, and to eat fresh fruits such as mango and papaya, along with reminders regarding taking medications and keeping medical appointments. Dr. Degani, our anthropologist colleague whom I mentioned in chapter 5, assisted us by making sure the app was appropriately adapted. According to my mentee Yvonne Commodore-Mensah, who was born in Ghana and is one of the study's two principal investigators, many people in Ghana own at least two mobile phones. "We want to take a device that is used frequently and see whether including this app in their daily lives will help control hypertension by having people measure their blood pressure every day and send those readings to their providers to allow them to track and monitor their progress," Dr. Commodore-Mensah said.

Using protocols developed for the RICH LIFE Project described in chapter 4, the ADHINCRA study seeks to apply chronic disease management interventions from health clinics to low-resource rural settings where people have non-Western attitudes about health and the body. Using community-based nurses who've been trained to monitor blood pressure, the clinical work is supplemented by education and behavioral reinforcement techniques designed for the unique cultural context of sub-Saharan Africa.

The study is based at four hospitals in western Ghana: Komfo Anokye Teaching Hospital, Kumasi South Hospital, Manhyia Government Hospital, and Suntreso Government Hospital in Kumasi, Ghana. The study places particular emphasis on patients from low-income and rural contexts. We have enrolled a total of 240 patients for a one-year period, with six months of active intervention by the clinical team followed by six months of tracking patient progress to assess outcomes.

Before launching the study, Dr. Commodore-Mensah, Mrs. Molello, and I traveled to Kumasi in January 2019 to meet with our academic and clinical site partners and with the leaders of the participating clinical sites. The study's co-principal investigator, Fred Stephen Sarfo, is a neurologist, researcher, and educator at KNUST. We had been introduced to him by another of my mentees, Linda Mobula, a general internist who currently works for the World Bank and who's been heavily involved in global chronic disease management efforts as well as the US State Department's public health emergencies management programs for the West African Ebola outbreak in 2014–2016, and the recent Ebola outbreak in the Democratic Republic of Congo. During our trip, we met with the deans of the School of Medical Sciences and the School of Public Health at KNUST, the Ashanti regional director at Ghana Health Services, and medical directors at our participating sites. We also met with representatives from Medtronic Labs who demonstrated the Akoma Pa app. We were able to view the app in action, determine whether it was suitable for the project, and to firm up our partnership. As Dr. Commodore-Mensah said, "I think the stars were aligned."

One of the themes we've heard over and over in our conversations with community partners in the United States and overseas is that academic institutions come into their communities to do studies, publish important papers and in some cases make financial deals with large pharmaceutical and biotechnology companies, and then leave without ever coming back to share the results or to help the communities to benefit. We wanted to establish strong and authentic relationships and let our partners know that we were in this for the long haul, not just for our own benefit. Because Dr. Commodore-Mensah is originally from Ghana and I'm originally from Liberia, we also have interests in adapting and disseminating interventions to reduce the burden of cardiovascular disease, with a focus on those most in need, to other sub-Saharan African countries. Having that cultural connection with our partners was helpful.

Six months later, in July 2019, Dr. Commodore-Mensah returned to Kumasi with two of the Center's trainees to train the local study staff and coordinate with the partnering study team at KNUST. One of our doctoral students, Kathryn Foti, said that being able to visit the hypertension clinics in person helped the team develop and refine the intervention protocol. "I think we had ideas about how the patient flow, the recruitment, the screening process, and the enrollment would work, so it was really valuable to be there physically and see the space—how everything was laid out," she said. "A good portion of our time was also spent ironing out the protocol with the staff who are going to be implementing it."

In the control group of the study, participants receive six months of enhanced usual care. Every day, these patients receive SMS messages that deal with healthy lifestyle behaviors, including smoking, diet, and physical activity but not with medication adherence or hypertension-specific issues. Every three days, they also receive an automated SMS directing them to a different short video clip about healthy living. Patients in this arm of the study receive usual care as determined by their providers.

In contrast, participants in the active intervention group use the Akoma Pa app to improve their communication with their community health officers. The use of this app by participants includes reminders, community health officer messaging, home blood pressure tracking, education materials (including education models on reducing the risk of cardiovascular disease and stroke that are tailored to the participant's knowledge level), and a community health officer provider portal that includes decision support tools. Even though the app had been developed in Ghana, we realized that some of the terminology and motivational messages it used might be unfamiliar to many who lived there. For example, in a list of side effects, one of the words used was "light-headedness," which is not a term that many Ghanaians use. So we changed the term to "feeling weak and dizzy." Even "headache" was an uncommon term in Ghana. Instead, we change the term to "a pain in my head."

OVERCOMING BARRIERS TO CARE

In Ghana, these mobile technologies can overcome some of the barriers to care that people face. Many people in Ghana live far from health care clinics and must travel for hours on unpaved roads in rural areas, or on paved roads that are congested with traffic due to overcrowding in urban areas. By the time people arrive at a clinic, they have to wait in long lines to see a doctor. Shortages of health care professionals pose further challenges to receiving adequate care.

By allowing people to access services remotely, the intervention seeks to mitigate some of these accessibility issues. For instance, the intervention allows the app to record blood pressure readings, which then can be sent to a community health officer. The officer can confer with a patient's doctor to adjust their medications and treatments without requiring the patient to come into the hospital.

The primary outcome measures of the study will be change in blood pressure levels from baseline, as well as change in blood pressure control by poverty status from baseline. In addition to analyzing changes in blood pressure and medication adherence, we'll look at how the patients responded to the technology.

Dr. Commodore-Mensah plans to examine whether patients continue to adhere to treatment and prescribed behavior after the active phase of the trial has ended. "We're going to leave the blood pressure monitors and smartphones with patients and see what they do and see how they're able to manage on their own," she said. "Hopefully at that point they should continue some of these self-management behaviors that the nurses have reinforced for the first six months."

In January 2020, Dr. Commodore-Mensah, Mrs. Molello, and I returned to Ghana to follow up on the ADHINCRA study. By this time, the program had achieved striking successes. The team had already enrolled 238 of the targeted 240 patients. Our academic and clinical partners were passionate about the program. Patients were measuring their blood pressure at home and sharing that information with their health care providers. We met patients who told us about how the program was helping them eat healthier, lose weight, and take their medications more regularly. Many were excited that they could use their mobile phones to ask questions and get advice from doctors and nurses. It saved them from worrying about their health, and it saved them time, because many of them lived far from a clinic and had to take buses or taxis every time they came in for a visit.

By controlling their blood pressure, we knew they would be less likely to have a stroke or heart attack or to develop kidney failure. Seeing people in the experimental group take control of their health and their lives was very gratifying, and preliminary data from home blood pressure monitors reveal encouraging improvements in that group's control rate.

COMMUNITY ADVISORY BOARDS ACROSS THE GLOBE

Our work with community advisory boards (CABs) in Africa is an example of how we're applying lessons learned in Baltimore to an international context and learning lessons from Africa that we might apply in Baltimore. In 2019, a team from the Center for Health Equity, including Reverend Debra Hickman, the co-chair of our community advisory board, and Mrs. Molello, traveled to Kampala, Uganda, to meet in person with CAB members to share advisory board experiences and contribute to a new evaluation tool assessing the effectiveness of community-academic partnership. During their visit, the team met with and studied three different CABs in place in Uganda. The first was coordinated by Baylor-Uganda and focuses on addressing the social stigma and prejudices that prevent sex workers, members of the LGBTQ community, and HIV-positive people from achieving optimal health. The second CAB that participated in the design sessions was Young Generation Alive, founded in 2005 by a group of young HIV patients to help other teenagers overcome the depression and fear that often accompany an HIV diagnosis. This CAB seeks to inspire and empower HIV-positive teens so that they can lead fulfilling lives despite their ongoing health challenges. The third participating CAB aims to ensure that children impacted by HIV receive psychological and social counseling that is appropriate for their age and development.

The team's efforts resulted in developing a three-part prototype digital toolkit designed to provide guidance to groups working with CABs. The first module of the toolkit focuses on capacity building, with information on how to design, form, and govern a CAB. The second module addresses the roles that CABs play in developing budgets and work plans for health equity interventions. A third module provides guidance on how CABs might most effectively conduct community outreach and disseminate research information. Each section of the toolkit includes performance checklists to assist with evaluating the various components of a CAB and its partnerships. "Our vision for the toolkit is for the local and global partners to create what they would find useful," said Mrs. Molello. "In the future, we'll test this toolkit in a larger and more diverse group of academic/community partnerships to obtain additional information about its acceptability and effectiveness."

TRANSFERABLE SOLUTIONS

In the fall of 2018, the Center for Health Equity and the Johns Hopkins Alliance for a Healthier World co-sponsored a workshop designed to launch new collaborations based on existing relationships. As part of our mutual missions to support innovative ways to address health inequities, the Alliance and the Center awarded several $10,000 grants to participating groups from the workshop that identified new areas to advance methods and practices in global-to-local work. "Although it's a modest amount of money, it can allow the teams to either fly (to Kenya, Uganda, or Ghana) or come back here; I think that has a big impact," Mrs. Molello said. "It allows the teams to work together on problems they're trying to solve in their different locations that could then have bigger influences in obtaining more funding and securing larger grants."

Another element in the Center's local-to-global and global-to-local approach is understanding that solutions created in resource-poor contexts also have the potential to improve care in resource-rich settings. For example, we know from designing the Center's interventions using community health workers that many similar strategies have been successful for addressing maternal and child health problems as well as infectious diseases in low-income countries across Asia and Africa. Resource-poor settings often allow people to be truly innovative with the tools they have, and this knowledge can be beneficial for developing efficient approaches in more developed countries. Resource-poor settings have even helped to inform mobile-technology solutions for health care applications. For example, the SMS texts many of us in the United States receive to remind us of a doctor's appointment or a prescription refill have been used in sub-Saharan Africa for many years with persons who are HIV positive or who have tuberculosis, as part of efforts to remind them when to take their medications and when they have appointments with care providers.

The use of innovative technologies also makes it possible to involve underserved communities at all stages of the design, implementation, and evaluation of interventions. The result will be health informatics and digital health interventions with a more informed awareness of the social context in which people actually live, learn, work, play, and pray.

RESPECTING LOCAL AND GLOBAL DIVERSITY

Low- and high-technology solutions can be applied broadly across countries. Researchers, educators, practitioners, community stakeholders, and leaders can all work together toward health equity on a local scale, across national and international boundaries, and in communities all over the world. Building these global networks will require respect for diverse experiences, perspectives, and policies, as well as flexibility and adaptability. Interventions must acknowledge history, uphold cultural values, and meet people where they are. This requires significant investments of time and frequent and clear communication.

Scientists and practitioners from developed countries often experience culture shock when they travel to parts of the world where the cultural and social mores may differ from their own, where political influences on health systems and human rights may differ, and where living conditions may be much worse and resources in medical facilities much more limited. On the other hand, they may be surprised to see the level of innovation and resourcefulness in the face of challenging circumstances. They're often humbled by the realization that despite the educational and economic advantages they may have, they are no more prepared to deal with the circumstances in communities impacted by health inequities than the people on the ground are. They also learn that all collaborators should be equal partners in decisions that inform medical and public health programs. Using feedback from local policymakers, practitioners, scientists, and laypersons, tailored approaches to care can be developed to address the social determinants of health.

As always, the best way to facilitate new partnerships is through a known member within a community. While I take similar approaches everywhere I work to address health equity, I'm often seen as an outsider in Ghana or Liberia because I don't live, teach, research, or work there on a regular basis. As a result, I usually rely on locals to introduce me to people and to show me what's expected. In Baltimore, it might be appropriate to show up at an event and begin talking with people right away. In Liberia or Ghana, if you arrive as a stranger to a community event, you're not expected to just start talking. Instead, a member of the community has to introduce you to everyone and explain why you're there. Furthermore, cultural norms extend beyond conversation—in Liberia, you might begin with a prayer or even by sharing a small meal. Such differences make it especially important to look, listen, and learn before asking for anything from the community.

I've been asked whether the motto of our international efforts is "think globally, act locally" or "think locally, act globally." Actually, it's both. Local results can inform global movements, and vice versa. Education, training, and capacity building can advance health equity in low- and middle-income countries as well as in high-income countries.

Some problems affect people no matter where they are in the world. It's another measure of the ways in which we're all connected. Due to urbanization, which has led to more sedentary lifestyles, consumption of processed foods, and more crowded living conditions, countries across the globe are experiencing a rise in the burden of noncommunicable diseases and conditions such as obesity, cardiovascular disease, diabetes, and cancer. Scientists are also documenting climate change's influences on the emergence and reemergence of infectious diseases—directly through climate's effect on infectious agent biology and distribution and on alterations in human immune systems, and indirectly through its effects on regional food supply and human migration patterns. Global travel patterns further enhance disease transmission and can lead to deadly pandemics, as we'll discuss in the next chapter. Thus, solving these global health concerns will require interdisciplinary partnerships among clinicians, biological, social, and environmental scientists, multisector leaders, and community members, across national and international boundaries.

CHAPTER 7

Health Equity in the Era of COVID-19

THE CORONAVIRUS PANDEMIC STARKLY ILLUSTRATED the adverse consequences of health disparities within the United States and around the world that have existed for generations, even centuries, within minority and other vulnerable populations. It also grimly underscored how such inequities impact individuals who might otherwise think that the health of others isn't their concern.

The virus first emerged in China in 2019. Despite efforts at containment, it quickly spread across Asia to Europe and the Western Hemisphere, carried primarily at first by wealthy travelers who could afford trips to distant destinations. After China, the next hot spot was in the Lombardy region of Italy, where news coverage again emphasized the area's relative wealth. Soon after, the virus spread to the United States, showing up first in long-term care facilities. But early reports of coronavirus infections of NBA stars also gave the impression that the disease was a condition for the rich and well traveled.

Not for long. Just months after the virus emerged, it became clear that people of color and those with low income were overrepresented among reported COVID-19 infections and deaths. In New York City, the predominantly ethnic minority neighborhoods of the Bronx had the most COVID-19 hospitalizations and deaths, while the predominantly White borough of Manhattan had the lowest rates—despite being more densely populated overall, although not necessarily within each home. This pattern was repeated across the country. At the end of March 2020, Michigan public health officials reported that African Americans accounted for 40 percent of confirmed cases and 40 percent of the state's deaths, despite constituting only 14 percent of its population.[1] In the first week of April 2020, slightly more than 70 percent of all the deaths in Louisiana were among African Americans, even though they account for just a third of the state's population. COVID-19 has also taken a disproportionate toll on many Indigenous communities in the United States and Latin America, among whom its full impact may never be known because of racial misclassification and the exclusion of Indigenous communities from data sets and analyses that are often used to make health policy decisions.

By February 22, 2021, the number of reported deaths from COVID-19 in the United States exceeded 500,000. During this same period, the American Public Media Lab reported that the overall mortality rates among Blacks, Latinos, Native Americans, and Pacific Islanders, adjusted for differences in the age distribution of these groups, were two to three times as high as the rates for Whites.[2]

Although these statistics are shocking, they're not surprising. The people most vulnerable to severe illness from the novel coronavirus are those with chronic medical conditions that are disproportionately found in socially disadvantaged populations due to entrenched structural and social inequities. Racial and ethnic minorities, low-income individuals, those living in more impoverished rural areas, people with serious mental illness, and those with disabilities all tend to develop illnesses such as heart disease, diabetes, and lung disease more frequently and at younger ages.

Furthermore, those most at risk tend to be those who are least able to protect themselves—and others—from infection. They often live in overcrowded apartments, use public transportation, and may need to shop more frequently for basic necessities such as food and medications because of limited and unstable access to funds. They're less likely to have health insurance or easy access to medical facilities and may rely on emergency care instead. In many places, emergency rooms discourage people from coming in for minor ailments, advising those with concerns to see their primary care physicians instead. However, this leaves residents in economically challenged neighborhoods who don't typically have primary care physicians without options. Left untreated, many of these non-COVID illnesses can also become life-threatening. These barriers are further magnified during a pandemic due to efforts to reduce the spread of infection and reserve health care services only for those who become infected.

Socially and economically disadvantaged populations are also overrepresented in essential jobs in transportation, government, utilities, health care, cleaning, and food supply services, and they're more likely to be employed in low-wage or temporary jobs that may not allow telework or provide sick leave. These are the people who ensure that we and our relatives can buy groceries, have uninterrupted water, power, and sanitation services, receive qualified and kind care, and so much more that we generally take for granted. Fear of lost wages or loss of employment may lead vulnerable members of society to try to work when they're ill, contributing to further spread of diseases within their communities. They may not be able to afford internet access and telephone services, placing them at greater risk of social isolation and of being uninformed about what they need to do to protect themselves and others. Due to historical and current experiences of discrimination and stigma, they may also distrust the very institutions they need to protect them during a pandemic.

Many are frightened of going to the doctor because they're uninsured and know that the bills will be astronomical, and undocumented residents fear being deported. They may live in senior housing, be disabled, or require assistance from others for day-to-day activities. They may even be homeless, making it virtually impossible to shelter in place.[3]

The coronavirus might also be more deadly for African Americans and other minority groups due to the consistently high levels of stress that existed long before its arrival and which it deepened. This stress, sometimes referred to as "weathering," results from lifelong exposure to discrimination and racism, which manifests in all areas of life and causes a kind of accelerated aging that's visible at a cellular level. Premature, stress-induced aging may make people in these communities more susceptible to the virus and other outbreaks, even among individuals who are younger than 65. When people are routinely disrespected, denied access to equitable education and health care, and are forced through employment barriers to live in places that are unsafe, the effects are long-lasting and severe, even if they're less easily or immediately observed. When other people see an unarmed Black man detained and choked to death by a White policeman while pleading that he can't breathe, the extent of anti-Black racism becomes obvious and draws the attention of national and global media, but many other inequities are less obvious and form a corrosive mass of cumulative disadvantages and obstacles.

Policy brutality is blatant manifestation of health inequities; in a 2013–15 national study, 49 percent or nearly 40,000 Black adult respondents reported being exposed to at least one or more police killings of unarmed Black individuals in their home state during the three months prior to completing the survey.[4] Black respondents experienced an average of 4.1 poor mental health days in the preceding month; for each additional police killing of an unarmed Black person during the preceding three months, Black respondents reported more poor mental health days. This effect was not observed among White respondents in the same states. One real concern is that the heightened attention to police violence perpetrated against people of color in 2020 will continue to have negative mental health spillover effects in the post-pandemic period. It's not surprising that given the collision of months of a pandemic and yet another wave of brutal police killings, a storm of outrage formed in 2020, leading many people across the country (of all races) to participate in Black Lives Matter protests. While the great majority of these protests were peaceful, a number of police forces used violent means such as tear gas and rubber bullets to break them up, which further underscores that what happens in one person's life and to their health extends to those far beyond their immediate circle.

The social determinants of health within vulnerable communities may be further exacerbated by the coronavirus pandemic. These determinants include struggling education systems (including, with remote schooling, unequal access to computers and wireless access to use them among students), scarce access to healthy food, and a dearth of places to safely get outside and exercise or enjoy nature. At the height of the pandemic, in many areas only supermarkets and pharmacies remained open. What if a person lives in a community that doesn't have either of these businesses and doesn't own a car to get to them? What will happen to people living with chronic conditions who don't have savings and who lose their jobs? Will they have to choose between food or medication? The coronavirus has forced many to make hard decisions, but people living in vulnerable communities have been the hardest hit. Many have had to make choices that may imperil their health and the health of others well beyond the coronavirus.

As the infection spread around the world, the saying "We're all in this together" was widely repeated across social media networks. However, this sentiment doesn't really get to the heart of the situation. Not all of us have a fair opportunity to avoid the disease or to recover from its effects. My colleague David R. Williams, an internationally recognized social scientist and professor of public health, sociology, and African American studies at Harvard, has said, "We're all in the same storm, but we're not all in the same boat."

EXCESS DEATHS AND COMMUNITY BEREAVEMENT

By the end of 2020, scientists projected that the total number of excess deaths in the United States that year in comparison to the previous years would exceed 400,000—all attributable to the COVID-19 pandemic. Although half of the deaths will be directly attributed to COVID-19, the rest will be linked to nonrespiratory complications of COVID-19 or societal disruptions that reduced or delayed access to health care and worsened other social determinants of health.[5] These excess deaths are disproportionately occurring among individuals living in the United States who are Black, Indigenous, Latino, or Pacific Islander—those with the highest per capita hospitalization and death rates. On October 13, 2020, the American Public Media Research Lab reported that if they'd died of COVID-19 at the same rate as White US residents, about 21,800 Black, 11,400 Latino, 750 Native American, and 65 Pacific Islander Americans would still be alive. A report published in February 2021 found that US life expectancy dropped by a whole year in the first half of 2020, and that the greatest drops were observed for Black men (3 years), Latino men (2.4 years), and Black women (2.3 years).[6]

This isn't a new phenomenon—communities of color have borne the burden of excess deaths from health disparities for generations. In an article published in the *Du Bois Review*, Mary Jackman and Kimberlee Shauman estimated that almost 7.7 million excess deaths occurred among Black individuals from 1900 to 1999.[7] Although excess deaths were highest during the early decades of the twentieth century, in subsequent decades they declined only modestly, and the number in the last decade of the twentieth century was almost as high as the level in the first decade. Over the course of the century, these deaths began to occur among older Black persons in the prime of life, with devastating effects on the economic and social well-being of their families and communities.

Racial disparities in life expectancy also mean that Black individuals in the United States are exposed to more family deaths than White individuals, from childhood through adulthood, exacerbating the aforementioned compounded stressors. In a nationally representative study published in the *Proceedings of the National Academy of Sciences*, Debra Umberson and colleagues estimated racial differences in exposure to the death of family members at different ages, beginning in childhood.[8] Their results confirmed that Blacks are significantly more likely than Whites to have experienced the death of a mother, a father, and/or a sibling from childhood through midlife. Additionally, between young adulthood and later life, Blacks were more likely than Whites to have experienced the death of a child and of a spouse. The authors suggested that more frequent and earlier exposure to family member deaths could contribute to cumulative health disadvantage across generations. The excess deaths of the COVID-19 pandemic could heighten these existing vulnerabilities among people of color, contributing to deep and prolonged "community bereavement."

Although lives lost can never be replaced, healing and renewal are possible for those who remain, through individual and societal acknowledgment of the harm created by centuries of injustice, commitments to rectifying past wrongs, and changes that restore all individuals and communities—but especially those who've lost the most—back to a state of health and wholeness.[9]

COVID-19 Morbidity and Mortality Statistics, April–December 2020

WORK FROM HOME RATES: Black and Latino Americans are overrepresented in low-wage jobs that offer the least flexibility and increase their risk of exposure to the coronavirus.

Source: U.S. Bureau of Labor Statistics, Job Flexibilities and Work Schedules — 2017–2018 Data from the American Time Use Survey, infections, hospitalizations and deaths, all by race/ethnicity.

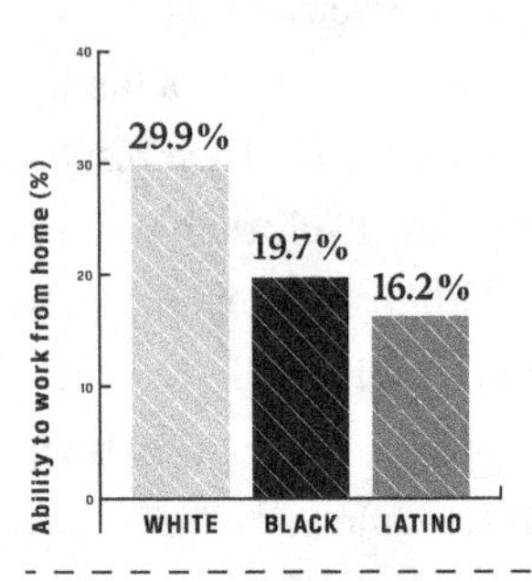

INFECTION RATES (per 100,000, through December 10, 2020): Black, Latino, and Native Americans are less likely to have access to quality housing, food, and water, and more likely to work in essential jobs that cannot be performed from home, placing them at greater risk of exposure to the novel coronavirus.

Source: https://covidtracking.com/race/infection-and-mortality-data

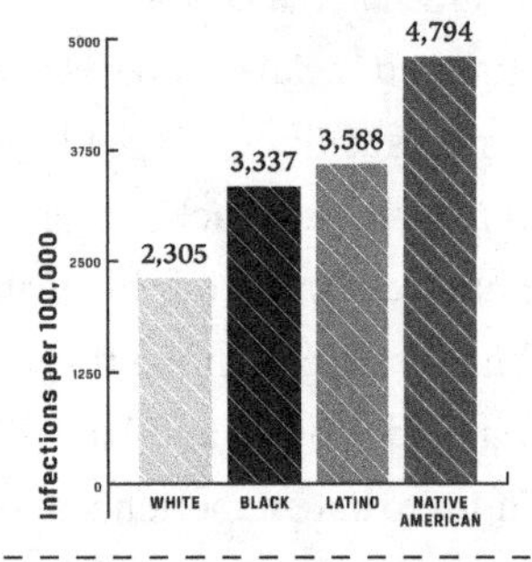

HOSPITALIZATIONS (per 100,000, March 1–November 28, 2020): Black, Latino, and Native Americans who contract the virus are more likely to suffer from pre-existing conditions that increase the risk of severe illness. Overrepresented among the uninsured, they may delay seeking treatment and may be sicker than White patients when they finally do.

Source: https://www.cdc.gov/coronavirus/2019-ncov/

Hospitalizations per 100,000
600
450
300
150
0
140
475.5
533.7
521
WHITE
BLACK
LATINO
NATIVE AMERICAN

AGE-ADJUSTED DEATHS (per 100,000, through December 8, 2020): From March to December 2020, Black, Latino, and Native Americans died of COVID-19 infections at rates that were 2.5 to 3.0 times the rates of White Americans when accounting for differences in the age distributions of these groups.

Source: APM Lab website: https://www.apmresearchlab.org/covid/deaths-by-race#rates

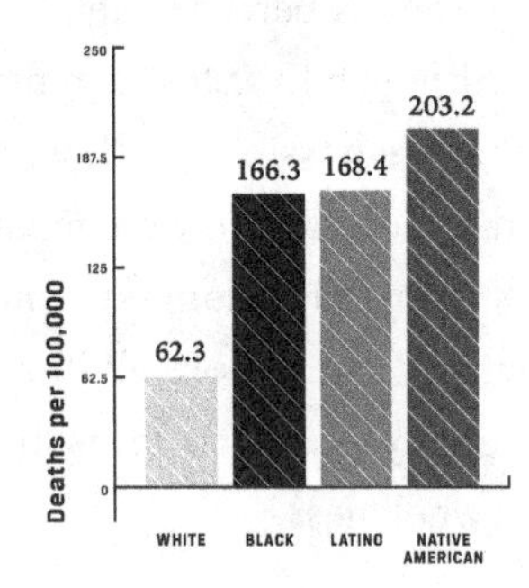

Source: Johns Hopkins Center for Health Equity

DISMANTLING THE INFLUENCES OF STRUCTURAL RACISM ON HEALTH

The COVID-19 pandemic and 2020 demonstrations reinforced my decades-long experiences witnessing disparities in health and health care in the United States. It clearly and tragically demonstrated the ways in which public policies influence health, both directly and indirectly. For that reason, considering the best response to this crisis demonstrates the range of policies that can address health disparities.

In the short term, Josh Sharfstein, a nationally recognized public health expert and vice dean for public health practice and community engagement at the John Hopkins Bloomberg School of Public Health, and I suggested that local and national leaders do five things:

1. track and monitor data on racial disparities in the impact of COVID-19 (a practice that has now begun in some municipalities, but is by no means standard or comprehensive);
2. provide better access to testing and medical care;
3. communicate in a trustworthy and respectful manner with communities of color;
4. encourage employers to provide protective equipment and improve working conditions for essential and frontline workers; and

5. address the immediate risks vulnerable groups face with respect to stable housing, food security, and digital access for education and health care.[10]

All these strategies are needed to ensure that disadvantaged populations don't continue to bear the greatest burden of this pandemic—or the next one.

However, in the longer term, the United States needs to engage in a much more comprehensive response to the impact of structural racism and health inequities on our culture, our institutions, and all our people, but especially people of color. Dr. Williams and I described three broad strategies to do this in 2019, before we or anyone else had heard of COVID-19.[11]

First, policymakers should establish "communities of opportunity" to minimize the adverse impacts of structural racism. This would mean creating communities that provide resources for early childhood development, have policies in place to reduce childhood poverty, provide work opportunities and income support for adults, and ensure healthy housing and neighborhood conditions.

Second, health care systems need new emphases on ensuring universal access to high-quality care. They also need to strengthen preventive and primary care, address patients' social needs as part of health care delivery, and diversify the health care workforce to more closely reflect the demographic composition of the patient population, working in concert with community leaders, academic institutions, other societal sectors, and policymakers.

Third, new research is needed to identify the best approaches to building political will and support in order to address social inequities in health. This could include initiatives to raise public awareness of the pervasiveness of health inequities and the connections between social factors and health. These initiatives need to build empathy and support for addressing disparities by more broadly and effectively telling the stories of the people whose lives have been affected by those inequities, as we've seen more often since the media coverage of police brutality in the recent killings of Black people.

In August 2020, I contributed to the 2020 National Urban League's State of Black America (SOBA) report. Part of an essay I wrote, titled "The Silver Lining in COVID-19's Dark Clouds," speaks to these recent experiences:

> The silver lining during these dark times is that this pandemic has revealed our shared vulnerability and our interconnectedness. Many people are beginning to see that when others don't have the opportunity to be healthy, all of us are at risk. I am hopeful because I see the pandemic producing a shift in thinking among many as they acknowledge the disparity between the lives of white people and people of color in this country. As we have seen through the Black Lives Matter protests, many

> people are finally recognizing the inequities that are borne out of systemic racism, becoming motivated to speak and act for justice and change. We know now, more than ever, that everyone's voice is important to bring about the change that we seek. The crises . . . are forcing us to confront the injustices and eliminate the inequities that prevent us from living up to our stated ideals of "liberty and justice for all."[12]

THE POWER OF PARTNERSHIPS IN PANDEMICS

Community-based organization partnerships (as discussed in chapter 4) need to be one of the strategic drivers of health promotion. With the coronavirus or any future pandemic, health and public health organizations must receive guidance and input on their initiatives directly from these organizations and from leaders who understand the specific needs and available resources for their most vulnerable citizens. Public health departments can work with these organizations to offer testing, conduct contact tracing, deliver education, provide supplies such as food and personal protective equipment, help people apply for unemployment, and other services. Public officials and health care leaders need to engage with these trusted messengers, including the leaders of faith communities, to provide support and relay important information about the pandemic to vulnerable populations as quickly as possible, including instruction on how people can protect themselves and others, and where and when to seek health care services.

Communicating this information is especially important to counter widespread misinformation among communities about the pandemic—or any global or even major regional health crisis. The spread of myths and misinformation is not unusual during a health crisis, and particularly from one that started abroad—recall the fears during the early years of the HIV/AIDS epidemic about contracting HIV from even doorknobs or casual physical contact with infected people. Furthermore, these myths tend to take on racial overtones, especially with a virus about which much remains unknown and unpredictable. Many Asian Americans have reported facing greater discrimination since the onset of COVID-19; these negative social attitudes have likely been fueled by people of influence calling it the "Chinese" virus. Another example, from the early days of the pandemic, is a widely spread myth that Black people were immune to the virus, which made many African Americans believe that they didn't need to protect themselves. With such rampant spread of misinformation, health care systems and public health departments have the responsibility to clearly communicate the dangers of the virus to racial and ethnic minority communities, getting messages to people where they are, through the best traditional and technological means available. City bus or subway messaging, for example, is a simple and effective means of delivering this information. Messages need to be accurate and include disclosures about what is and is not known and about the difference between myths and truths. They need to take into account specific concerns and fears while conveying empathy and respect for the wisdom and experience that exist within these communities.

In addition to the concern that mistrust and skepticism might prevent people of color from seeking health care when they need it is the risk that it will also lead to a reluctance to participate in clinical trials of vaccines and treatments for COVID-19, and a refusal to accept these treatments and vaccines even once they are widely available. Thus, it is critical that health care professionals and scientists work to enhance their trustworthiness through the relationship-centered and structurally competent approaches described in chapter 3. In recognition of the critical importance of this issue, in September 2020, the National Institutes of Health announced a $12 million initiative for outreach and engagement efforts in ethnic and racial minority communities disproportionately affected by the COVID-19 pandemic, called Community Engagement Alliance Against COVID-19, on which I serve as steering committee co-chair.[13]

People in positions of power or influence need to provide leadership on specific problems and appropriate solutions. For instance, after testing positive for COVID-19, Idris Elba, a renowned British actor of African descent, went online to quash the myth that only White people could get the virus. He wrote:

"Black people, please, please, please, please understand that coronavirus, you can get it. There are so many stupid, ridiculous conspiracy theories about Black people not being able to get it. That's dumb, stupid. All right? That is the quickest way to get more Black people killed."[14] Elba's words resounded with people across many nations. Everyday people who have survived COVID-19 or who have enrolled in clinical trials testing treatments or vaccines can lend their voices to engage communities of color in safe practices, self-advocacy to protect themselves, ways to get the help they deserve, and ways to contribute to research.

GREATER AWARENESS OF INEQUITIES

Our entire population needs greater awareness of the existence of health disparities and of the actions that can be taken to reduce these disparities. A 2010 survey by the US Department of Health and Human Services' Office of Minority Health and NORC at the University of Chicago of 3,159 adults ages 18 or older found that 59 percent of Americans were aware of racial and ethnic health disparities that disproportionately affect African Americans and Latinos. This finding demonstrates only a modest increase from the 55 percent who reported awareness of such health disparities in a 1999 survey by the Kaiser Family Foundation. In the 2010 survey, 89 percent of African American respondents were aware of African American and White disparities, versus only 55 percent of Whites.[15] These and other survey results reveal the importance of implementing health promotion strategies that raise awareness of health disparities. Without the best available knowledge, without knowing what your community does or doesn't have access to, it's hard to act in your own best interest. Patients have to be aware of health care disparities in order to self-advocate for just treatment.

The health care workforce in particular needs to know more about health disparities. In a recent survey of 115 primary care providers from two large academic centers in Colorado, questions assessed provider recognition and perceived contributors of disparities in hypertension control.[16] Among respondents, 86 percent agreed that disparities in race and ethnicity existed in hypertension care, and 89 percent agreed that disparities in socioeconomic status existed in hypertension care within the US health system. However, only 33 percent and 44 percent thought racial and ethnic and socioeconomic disparities existed in the care of their own patients. Therefore, despite growing acknowledgment of the impact of quality of care on health disparities, a disconnect remains between awareness and practice, particularly the ways in which even more sensitive providers continue to contribute to these disparities.

Community partnerships can also foster greater awareness of health disparities by engaging a diverse array of organizations, including health and human service institutions, nonprofit agencies, government, businesses, educational institutions, and community- and faith-based institutions. These partnerships reach a wide range of populations. Baltimoreans United in Leadership Development (BUILD), for example, an interfaith, multiracial community power organization, is a partnership between congregations, public schools, and neighborhood associations from across Baltimore that has worked to improve housing, increase job opportunities, rebuild schools and neighborhoods, and enhance access to mental health and substance abuse treatment, among other issues, for over 40 years. Since the onset of the COVID-19 pandemic, BUILD has partnered with Johns Hopkins University to provide food relief and personal protective equipment to more than 1,200 families. The organization has also partnered with medical institutions and the city health department to provide community-based COVID-19 testing. Organizational networks, media outlets, innovative uses of information technology, and educational approaches can all be used to significantly increase awareness of health inequities and inspire greater involvement in finding and implementing solutions.

Since the onset of the COVID-19 pandemic, I've had the opportunity to share this information directly with those who have power and influence by providing briefings to numerous policymakers, including a closed-door session with the Democratic senators in the US Congress, and a special session with mayors across the United States and around the world. I've also participated in briefings with Baltimore City Council members and Maryland county executives. I've had the opportunity to inform and discuss these issues with business leaders, educational institutions, and museums, including the Smithsonian Institution's National Museum of African American History and Culture. I've spoken to national civic organizations, including historically Black fraternities and sororities, and with journalists and reporters from major local, national, and international news outlets, including NPR, CNN, Politico, PBS NewsHour, the BBC, the *Baltimore Sun*, *U.S. News and World Report*, the *New York Times*, and *The Guardian*. I've spoken to faith communities as well as a global audience of women of color through the Essence Wellness House online platform, and to more general minority populations through the National Urban League, as mentioned previously. On the one hand, it's been challenging to talk about the incredible suffering and loss of life endured by communities of color during this time, and to manage the time needed to respond to so many media requests while directing two programs at Hopkins. However, I've also been gratified to see so many people from different backgrounds, professions, and positions of power take an interest in health inequities and how to address them. This acknowledgment, while prompted by a global tragedy, gives me hope that the change we've needed for so long may finally come.

One of the most interesting opportunities I had to provide education on both health equity and pandemics was through contributing answers to questions on social dynamics and social behaviors to a new guide for youth ages 8–17 about life during the pandemic, developed by the Smithsonian Science and Education

Center in collaboration with the World Health Organization and the Inter-Academy Partnership. The guide aims to help young people understand the science of COVID-19 and take actions to keep themselves, their families, and their communities safe. It's now available in 21 languages and is being distributed through community partners across the United States and the globe.

Health care system and health equity research efforts should involve a variety of platforms—from print and broadcast media to social media to forums, meetings, marches, movements, and toolkits. During the COVID-19 crisis, I've been able to significantly broaden the reach of my team's findings and introduce new programs to local, national, and global audiences. I'm energized by the opportunity to build on relationships with colleagues, community advocates, national and international policymakers, and national media outlets to further address injustices, clarify misconceptions, and develop and scale innovative, community-engaged approaches. Each new panel discussion, briefing with politicians, article, and interview sparks new solutions to consider and ways to communicate critical messages during a time of extreme stress for everyone but disproportionately for minority and other vulnerable communities.

A NEW KIND OF "HERD IMMUNITY"

Social inequities took root because of choices made to value people from diverse backgrounds differently. Policies and systems dating back centuries have led us to where we are today. If we want to undo that, we have to look at the underlying structures, policies, and other factors that need to be changed. If we don't address all the necessary levels, we won't resolve the problem.

We still have a lot of work to do to achieve health equity. Nonetheless, awareness has increased over the course of my career. In 2020, we've seen a new level of social consciousness. We've returned to the journey toward a more just society that we'd strayed from since the civil rights movement of the 1960s.

In a 2020 Viewpoint article in the *Journal of the American Medical Association*, David R. Williams and I argued:

> The striking racial/ethnic disparities reported for COVID-19 infection, testing, and disease burden are a clear reminder that failure to protect the most vulnerable members of society not only harms them but also increases the risk of spread of the virus, with devastating health and economic consequences for all. COVID-19 disparities are not the fault of those who are experiencing them, but rather reflect social policies and systems that create health disparities in good times and inflate them in a crisis. The US must develop a new kind of "herd immunity," whereby resistance to the spread of poor health in the population occurs when a sufficiently high proportion of individuals, across all racial, ethnic, and social class groups, are protected from and thus "immune" to negative social determinants.[17]

PATHWAYS TO CHANGE

Health care disparities are crippling our health care system, our economy, our society, and our health. Simply put, as Margaret Chan, director-general of the World Health Organization, has said, "Health inequity really is a matter of life and death."[18] We can improve the health of the populations we serve, achieve health equity, and enhance the credibility of the health care system as a valued community partner. We can use rigorous research findings and methods to create new health care delivery models, develop model educational programs, and engage with the public in working for health equity. But success cannot occur without an all-hands-on-deck approach.

Despite the clear moral argument for eliminating health disparities, many health systems are not fully committed to the task. In part, this lack of commitment stems from uncertainty in how to address health inequities, including the fear of economic consequences. However, the transition from our current fee-for-service reimbursement system to value-based care will provide many opportunities to eliminate health disparities. Value-based systems used by payers, such as the Centers for Medicare & Medicaid Services and private insurers, reward health care providers with payments for the quality of care they give to people. In other words, these systems reward the good health outcomes that communities seek. Reimbursements based on value will incentivize health systems to mitigate disparities, as opposed to the traditional health care delivery model, which rewards the quantity, rather than the quality of health care and largely overlooks health-related social needs.

Achieving health equity will require reshaping the health care system to prevent adverse health outcomes among underserved populations rather than treating problems once they've arisen. Health systems must move beyond their traditional roles of providing high-quality health care to also promoting equity. They must partner with other sectors such as nutrition, housing, and transportation, commit to a shared vision of eliminating health disparities, and build on the strengths of each other's programs and successes.

Tackling health disparities will also require identifying the strengths, weaknesses, and resources available within communities and among a wide range of stakeholders. Achieving health equity will require that resources, talent, and ideas be pooled from many sources. To take just one example, simply encouraging people to eat healthier foods will not be enough to eliminate diet-related inequities. Grocery stores need to provide people with healthier options at more affordable prices. More restaurants need to display the caloric and fat content of their menu items so that people can make informed choices around what they eat. Schools need to provide healthy lunches, regardless of financial hardships. Food production companies need to restrict unhealthy food marketing practices, especially to children. Sea-change examples such as Chile's successful front-of-food-package warning labels for excess salt, fat, and sugar spearheaded by Dr. Ricardo Uauy, and refinement of the McDonald's Happy Meal to include more healthful items such as apple slices and the removal of soda from those menus (resulting in a 15 percent increase between 2013 and 2018 in the number of Happy Meals served with milk, water, or juice drink) illustrate that change at the uppermost governmental and commercial levels is possible. Still, policies directed at the entire population should consider the impact on different social groups, including the most vulnerable, and make adjustments as appropriate to avoid unintended harm.

Our goals need to be foremost in reframing the problem. As I wrote at the beginning of this book, many people think that getting rid of disparities requires giving everyone the same thing. In fact, achieving equity will require providing different things for different people and different groups. Determining which resources are needed and where will require a thorough understanding of the ways in which health inequities manifest across communities as well as the features shared among communities. At the Center, we're learning more about these complexities every day, and we're eager to share this knowledge widely.

That so many factors combine to target vulnerable groups can be overwhelming. Yet everyone's health is connected, both within communities and across the globe. Inequities in opportunities to live a healthy life also lead to other societal ills, including violence and civil unrest, which can lead to economic and political instability and poorer quality of life for all within a society. I have experienced this on a deeply personal level, in Liberia, where I grew up, and now, in the United States. We witness this phenomenon in news reports from across the globe. Societies with more inequity have shorter lifespans and poorer quality of life overall than do countries that have more equity, and these poorer outcomes are observed even among the most advantaged groups in a society that is marked by inequities. In his book *The Impact of Inequality: How to Make Sick Societies Healthier*, Richard Wilkinson observes that "the pathway runs from inequality, through its effects on social relations and the problems of low social status and family functioning, to its impact on stress and health."[19] Once people realize that poor health among the members of any group can affect the health of everyone, they will see that health inequities are not someone else's problem—they're everyone's problem. By acknowledging and acting upon the interconnectedness of our lives, everyone can help create a healthier and safer world. The late Representative Elijah E. Cummings, of Maryland's Seventh District, aptly said:

> If we want the nation to be strong, then the people have to be healthy; they have to be well. If there's anything that we can do to stamp out disparities, we need to do it, by any means necessary. One of the things that's so important is that people have, what I call, a liberated future. And it's hard to be liberated when you're not healthy.

> And when I say liberated, I mean freedom to be all that God meant for you to be. When we don't deal with disparities, then what we're doing is denying people the opportunity to give back to the world. So we've got to fight it. We've got to fight it with everything we've got.

In a similar vein, the late Congressman John Lewis, who served in the US House of Representatives from 1987 until his death in 2020, famously said, we have to "get in trouble, good trouble, to save the soul of this nation."

I firmly believe that everyone deserves the opportunity to live a healthy life. And when everyone is given that opportunity, we all benefit. This is why I continue my work to advance health equity through research, education, and advocacy, in collaboration with colleagues from diverse disciplines and in partnership with leaders from various sectors of society and members of impacted communities both here and overseas. At first, my community was Liberia; over the years, it expanded to Geneva, Switzerland, then Atlanta, then Baltimore, then to the rest of the United States, then back to Africa and places throughout the world. Today, I consider myself a global citizen. I've become attached to many of these neighborhoods, cities, and nations through my work and personal life, and have sought to find and create ways give back to those in need, wherever they live.

I'd always hoped that one day my research would uncover significant results that would draw the attention of professional societies, health system leaders, and policymakers and bring about real changes in practice and policy to those who needed them most. Honestly, I'm grateful for the influence it's had so far and I still hope for more of this. But I've also discovered that making a difference in the "real" world beyond academia requires going well beyond the publication of scientifically compelling results. A researcher might discover a new drug, but unless that drug is used, nothing will change. As my experience with relationships has expanded from local communities to the national and international level, I've gained an ever-greater appreciation of how each individual and each relationship is important, and how interdependent they all are. I've grown to appreciate that the rewards of these connections are just as gratifying as the most successful research projects.

Like so many, I'm inspired by the words of the Reverend Martin Luther King Jr., who said: "All labor that uplifts humanity has dignity and importance and should be undertaken with painstaking excellence. . . . Make a career of humanity. Commit yourself to the noble struggle for equal rights. You will make a great person of yourself, a greater nation of your country, and a finer world to live in." Dr. King elaborated on this sentiment in his April 12, 1963, letter from the Birmingham Jail, which was written on the day I was born. "Injustice anywhere is a threat to justice everywhere." My hope is that by reading this book, you'll understand why health disparities impact you and others around you, and that your expanded awareness will inspire you to support changes. I envision a future when everyone will have the ability to achieve strong health and pursue their dreams, as I've had the good fortune to do.

Acknowledgments

I AM GRATEFUL, FIRST AND FOREMOST, TO GOD, with whom all things are possible. I feel blessed beyond belief to have the opportunity to experience a healthy life, loving and nurturing relationships, meaningful work, and the ability to give back to my community and to society.

Thank you to the Bloomberg Distinguished Professors program, including its visionary benefactor, Michael R. Bloomberg, along with Denis Wirtz and Julie Messersmith, for providing me with this opportunity to share my experiences and scholarly work on an issue that is core to my very existence—that all people, regardless of background, deserve a chance to be healthy and live up to their highest potentials. Thanks as well to the entire team at Johns Hopkins University Press, including my editor, Anna Marlis Burgard, for her countless hours of reading, edits, and conversations, and to Steve Olson and Sarah Olson for their contributions.

For their steadfast love and support I thank my parents, Izetta Roberts Cooper and Henry Nehemiah Cooper, MD, who were my first and most important role models through their quests for professional excellence, passionate dedication to serving humanity, and their love of family. I'm blessed that my mother is alive to share this accomplishment with me. I'd also like to thank my brother, Armah Jamale Cooper, MD, and my sister, Dawn Cooper Barnes, PhD, my two closest friends in life. My brother modeled courage, unworldliness, and humor, and my sister continues to model selflessness, optimism, and a zest for life. I'd also like to thank my brother-in-law, Nathaniel Barnes, my nephews and nieces, extended family, close friends and colleagues and mentors, staff, community partners, patients, research participants, my pastor, the Reverend Mary Ka Kanahan, and church members at Saint John United Methodist–Presbyterian Church USA, who've been with me through the highs and lows of this journey, and whose faith and strength have inspired me. To my son Donovan, thank you for your understanding and encouragement, and for showing me the importance of perseverance and sharing in my failures and triumphs. And to my son Devin, although I never saw you grow up, your brief and beautiful life revealed courage within me and love around me that I never knew existed. Last, but certainly not least, to my husband, Nigel Green, my rock. Thank you for being my life partner, supporting me through every professional and personal challenge, making sure I make time for self-care and laughter, and helping me become a better person every day.

The African proverb "If you want to go fast, go alone. If you want to go far, go together" expresses a theme—the strengths of community and collaboration—that's interwoven throughout this book. With the support of all the people mentioned here, I've aimed to use my life experiences and my training as a physician, and a public health, social, and behavioral scientist, to bring awareness, empathy, hope, and healing to others. This book shares a portion of that story.

Notes

PREFACE

1. David R. Williams and Lisa A. Cooper. 2020. COVID-19 and health equity—A new kind of "herd immunity." *Journal of the American Medical Association* 323(24): 2478–2480.

INTRODUCTION. FIVE MILES APART

1. Paula Braveman. 2014. What are health disparities and health equity? We need to be clear. *Public Health Reports* 129(Suppl. 2): 5–8.
2. US Department of Health and Human Services. 1985. *Report of the Secretary's Task Force on Black and Minority Health*. Washington, DC: Department of Health and Human Services.
3. Alastair M. Gray. 1982. Inequalities in health. The Black Report: A summary and comment. *International Journal of Health Services* 12(3): 349–380.
4. Michaela Benzeval, Ken Judge, and Margaret Whitehead, eds. 1995. *Tackling Inequalities in Health: An Agenda for Action*. London: SAGE Publications, 1995.
5. Commission on Social Determinants of Health. 2008. *Closing the Gap in a Generation: Health Equity through Action on the Social Determinants of Health*. Geneva: World Health Organization.
6. World Health Organization, Health topics: Equity. https://www.who.int/healthsystems/topics/equity/en/.
7. David Satcher and George Rust. 2006. Achieving health equity in America. *Ethnicity & Disease* 16 (Suppl. 3): S3-8–S3-13.

CHAPTER 1. THE PERSONHOOD OF PATIENTS

1. Somnath Saha, Mary Catherine Beach, and Lisa A. Cooper. 2008. Patient centeredness, cultural competence and healthcare quality. *Journal of the National Medical Association* 100(11): 1275–1285.
2. M. Stewart, J.B. Brown, W.W. Weston, and T.R. Freeman. 2003. *Patient-Centred Medicine: Transforming the Clinical Method.* 2nd ed. Abingdon, TX: Radcliffe Medical Press; N. Mead and P. Bower. 2000. Patient-centredness: A conceptual framework and review of the empirical literature. *Social Science & Medicine* 51(7): 1087–1110; Rita B. Ardito and Daniela Rabellino. 2011. Therapeutic alliance and outcome of psychotherapy: Historical excursus, measurements, and prospects for research. *Frontiers in Psychology* 2: 270.
3. Karen Davis. 1999. Improving lives through information. *Health Affairs* 18(2): 219–225.
4. Lisa Cooper-Patrick, Neil R. Powe, Millie W. Jenckes, Junius J. Gonzales, David M. Levine, and Daniel E. Ford. 1997. Identification of patient attitudes and references regarding treatment of depression. *Journal of General Internal Medicine* 12(7): 431–438.
5. Cooper-Patrick, Powe, Jenckes, Gonzales, Levine, and Ford. 1997.
6. Lisa A. Cooper, Charlotte Brown, Hong Thi Vu, Daniel E. Ford, and Neil R. Powe. 2001. How important is intrinsic spirituality in depression care? A comparison of white and African-American primary care patients. *Journal of General Internal Medicine* 16(9): 634–638; Lisa A. Cooper, Junius J. Gonzales, Joseph J. Gallo, Kathryn M. Rost, Lisa S. Meredith, Lisa V. Rubenstein, Nae-Yuh Wang, and Daniel E. Ford. 2003. The acceptability of treatment for depression among African-American, Hispanic, and white primary care patients. *Medical Care* 41(4): 479–489; Jane L. Givens, Thomas K. Houston, Benjamin W. Van Voorhees, Daniel E. Ford, and Lisa A Cooper. 2007. Ethnicity and preferences for depression treatment. *General Hospital Psychiatry* 29(3): 182–191.

7. Andrew K. Schnackenberg and Edward C. Tomlinson. 2016. Organizational transparency: A new perspective on managing trust in organization-stakeholder relationships. *Journal of Management* 42(7): 1784–1810.
8. E. Moy, E. Chang, M. Barrett, and Centers for Disease Control and Prevention (CDC). 2013. Potentially preventable hospitalizations: United States, 2001–2009. *MMWR Supp*. 62(3): 139–143.
9. L. Ebony Boulware, Lisa A. Cooper, Lloyd E. Ratner, Thomas A. LaVeist, and Neil R. Powe. 2003. Race and trust in the health care system. *Public Health Reports* 118(4): 358–365.
10. Charles Blow. December 6, 2020. How Black people learned not to trust. *New York Times;* Michele Norris. December 9, 2020. Black people are justifiably wary of a vaccine: Their trust must be earned. *Washington Post.*
11. National Institutes of Health, news release. August 7, 2013. NIH, Lacks family reach understanding to share genomic data of HeLa cells, https://www.nih.gov/news-events/news-releases/nih-lacks-family-reach-understanding-share-genomic-data-hela-cells.
12. Lisa A. Cooper, Mary Catherine Beach, Rachel L. Johnson, and Thomas S. Inui. 2006. Delving below the surface: Understanding how race and ethnicity influence relationships in health care. *Journal of General Internal Medicine* 21: S21–S27.
13. Cooper, Beach, Johnson, and Inui. 2006.

CHAPTER 2. THE PATIENT-PHYSICIAN RELATIONSHIP

1. Mack Lipkin, Jr., Timothy E. Quill, and Rudolph J. Napodano. 1984. The medical interview: A core curriculum for residencies in internal medicine. *Annals of Internal Medicine* 100: 277–284.

2. Lisa Cooper-Patrick, Joseph J. Gallo, Junius J. Gonzales, Hong Thi Vu, Neil R. Powe, Christine Nelson, Daniel E. Ford. 1999. Race, gender, and partnership in the patient-physician relationship. *Journal of the American Medical Association* 282(6): 583–589.
3. Rachel L. Johnson, Debra Roter, Neil R. Powe, and Lisa A. Cooper. 2004. Patient race/ethnicity and quality of patient–physician communication during medical visits. *American Journal of Public Health* 94(12): 2084–2090.
4. Kimberly D. Martin, Debra L. Roter, Mary C. Beach, Kathryn A. Carson, and Lisa A. Cooper. 2013. Physician communication behaviors and trust among black and white patients with hypertension. *Medical Care* 51(2): 151–157.
5. Crystal W. Cené, Debra Roter, Kathryn A. Carson, Edgar R. Miller III, and Lisa A. Cooper. 2009. The effect of patient race and blood pressure control on patient-physician communication. *Journal of General Internal Medicine.* 24(9): 1057–1064; Bri K. Ghods, Debra L. Roter, Daniel E. Ford, Susan Larson, Jose J. Arbelaez, and Lisa A. Cooper. 2008. Patient-physician communication in the primary care visits of African Americans and whites with depression. *Journal of General Internal Medicine* 23(5): 600–606; Anika L. Hines, Debra Roter, Bri K. Ghods Dinoso, Kathryn A. Carson, Gail L. Daumit, and Lisa A. Cooper. 2018. Informed and patient-centered decision-making in the primary care visits of African Americans with depression. *Patient Education & Counseling* 101(2): 233-240.
6. Lisa A. Cooper, Debra L. Roter, Rachel L. Johnson, Daniel E. Ford, Donald M. Steinwachs, and Neil R. Powe. 2003. Patient-centered communication, ratings of care, and concordance of patient and physician race. *Annals of Internal Medicine* 139(11): 907–915.
7. Rachel L. Johnson Thornton, Neil R. Powe, Debra Roter, and Lisa A. Cooper. 2011. Patient-physician social concordance, medical visit communication and patients' perceptions of health care quality. *Patient Education and Counseling* 85(3): e201–e208.

8. Richard L. Street, Kimberly J. O'Malley, Lisa A. Cooper, and Paul Haidet. 2008. Understanding concordance in patient-physician relationships: personal and ethnic dimensions of shared identity. *Annals of Family Medicine* 6(3): 198–205.
9. Institute of Medicine. 2001. *Crossing The Quality Chasm: A New Health System for the 21st Century.* Washington, DC: National Academies Press; Institute of Medicine. 2003. *Unequal Treatment: Confronting Racial and Ethnic Disparities in Health Care.* Washington, DC: National Academies Press.
10. Mary Catherine Beach, Thomas Inui, and the Relationship-Centered Care Research Network. 2006. Relationship-centered care: A constructive reframing. *Journal of General Internal Medicine* 21 Suppl. 1: S3–S8.
11. Michelle van Ryn and Jane Burke. 2000. The effect of patient race and socio-economic status on physicians' perceptions of patients. *Social Science and Medicine* 50(6): 813–828.
12. Michelle van Ryn, Diana Burgess, Jennifer Malat, and Joan Griffin. 2006. Physicians' perceptions of patients' social and behavioral characteristics and race disparities in treatment recommendations for men with coronary artery disease. *American Journal of Public Health* 96(2): 351–357.
13. Lisa A. Cooper, Debra L. Roter, Kathryn A. Carson, Mary Catherine Beach, Janice A. Sabin, Anthony G. Greenwald, and Thomas S. Inui. 2012. The associations of clinicians' implicit attitudes about race with medical visit communication and patient ratings of interpersonal care. *American Journal of Public Health* 102(5): 979–987.
14. Alexander R. Green, Dana R. Carney, Daniel J. Pallin, Long H. Ngo, Kristal L. Raymond, Lisa I. Iezzoni, and Mahzarin R. Banaji. 2007. Implicit bias among physicians and its prediction of thrombolysis decisions for black and white patients. *Journal of General Internal Medicine* 22(9): 1231–1238.

15. Michelle van Ryn, Rachel Hardeman, Sean M. Phelan, Diana J. Burgess, John F. Dovidio, Jeph Herrin, Sara E. Burke, David B. Nelson, Sylvia Perry, Mark Yeazel, and Julia M. Przedworski. 2015. Medical school experiences associated with change in implicit racial bias among 3547 students: A medical student CHANGES Study report. *Journal of General Internal Medicine* 30(12): 1748–1756.
16. Lisa Cooper. August 28, 2018. Doctors can fight unconscious racial bias. *Bloomberg Opinion*, https://www.bloomberg.com/opinion/articles/2018-08-28/doctors-can-fight-implicit-bias-against-african-american-patients.

CHAPTER 3. DEVELOPING SOLUTIONS TO HEALTH DISPARITIES

1. E.L. Rosenthal, N. Wiggins, J.N. Brownstein, S. Johnson, I.A. Borbon, and R. Rael. 1998. *Final Report of the National Community Health Advisor Study*. Tucson: University of Arizona, Mel and Enid Zuckerman College of Public Health, Center for Rural Health, https://crh.arizona.edu/publications/studies-reports/cha.
2. American Public Health Association, Community Health Worker Section. https://www.apha.org/apha-communities/member-sections/community-health-workers.
3. Debra L. Roter. 2000. The medical visit context of treatment decision-making and the therapeutic relationship. *Health Expectations* 3: 17–25.
4. Sherrie H. Kaplan, Sheldon Greenfield, and John E. Ware Jr. 1989. Assessing the effects of physician-patient interactions on the outcomes of chronic disease. *Medical Care* 27: S110–S127; Debra L. Roter. 1977. Patient participation in the patient-provider interaction: The effects of patient question asking on the quality of interaction, satisfaction and compliance. *Health Education Monographs* 50:281–315.

5. Lisa A. Cooper, Debra L. Roter, Kathryn A. Carson, Lee R. Bone, Susan M. Larson, Edgar R. Miller III, Michael S. Barr, and David M. Levine. 2011. A randomized trial to improve patient-centered care and hypertension control in underserved primary care patients. *Journal of General Internal Medicine* 26(11):1297–1304.
6. Chidinma Ibe, Janice Bowie, Debra Roter, Kathryn A. Carson, Bone Lee, Dwyan Monroe, and Lisa A. Cooper. 2017. Intensity of exposure to a patient activation intervention and patient engagement in medical visit communication. *Patient Education and Counseling* 100(7): 1258–1267.
7. Lisa A. Cooper, Bri K. Ghods Dinoso, Daniel E. Ford, Debra L. Roter, Annelle B. Primm, Susan M. Larson, James M. Gill, Gary J. Noronha, Elias K. Shaya, and Nae-Yuh Wang. 2013. Comparative effectiveness of standard versus patient-centered collaborative care interventions for depression among African Americans in primary care settings: The BRIDGE Study. *Health Services Research* 48(1): 150–174.
8. Mary Catherine Beach, Eboni G. Price, Tiffany L. Gary, Karen A. Robinson, Aysegul Gozu, Ana Palacio, Carole Smarth, Mollie W. Jenckes, Carolyn Feuerstein, Eric B. Bass, Neil R. Powe, and Lisa A. Cooper. 2005. Cultural competence: A systematic review of health care provider educational interventions. *Medical Care* 43(4): 356–373.
9. Melanie Tervalon and Jann Murray-Garcia. 1998. Cultural humility versus cultural competence: A critical distinction in defining physician training outcomes in multicultural education. *Journal of Health Care for the Poor and Underserved* 9(2): 117–125.
10. Somnath Saha, Mary Catherine Beach, and Lisa A. Cooper. 2008. Patient centeredness, cultural competence and healthcare quality. *Journal of the National Medical Association* 100(11): 1275–1285.
11. David R. Williams and Lisa A. Cooper. 2019. Reducing racial inequities in health: Using what we already know to take action. *International Journal of Environmental Research and Public Health* 16(4): 606.

12. American Board of Internal Medicine Foundation / Timothy Lynch. 2018. [Re]Building trust: summary paper, https://abimfoundation.org/wp-content/uploads/2018/10/2018-ABIM-Foundation-Forum-Summary.pdf; American Board of Internal Medicine Foundation. October 18, 2018. Excerpts from the 2018 ABIM Foundation Forum: Lisa Cooper, https://abimfoundation.org/video/excerpts-from-the-2018-abim-foundation-forum-lisa-cooper.
13. Donald E. Wesson, Catherine R. Lucey, and Lisa A. Cooper. 2019. Building trust in health systems to eliminate health disparities. *Journal of the American Medical Association* 322(2): 111–112.

CHAPTER 4. THE JOHNS HOPKINS CENTER FOR HEALTH EQUITY

1. Lisa A. Cooper, Mary Catherine Beach, Rachel L. Johnson, and Thomas S. Inui. 2006. Delving below the surface: Understanding how race and ethnicity influence relationships in health care. *Journal of General Internal Medicine* 21: S21–S27.
2. Meera Viswanathan, Alice Ammerman, Eugenia Eng, Gerald Gartlehner, Kathleen N. Lohr, Derek Griffith, Scott Rhodes, Carmen Samuel-Hodge, Siobhan Maty, Linda Lux, Lucille Webb, Sonya F. Sutton, Tammeka Swinson, Anne Jackman, and Lynn Whitener. 2004. *Community-based Participatory Research: Assessing the Evidence.* Evidence Reports/Technology Assessments, No. 99. Rockville, MD: Agency for Healthcare Research and Quality.
3. Karen Fortuna, Paul Barr, Carly M. Goldstein, Robert Walker, LaPrincess Brewer, Alex Zagaria, and Stephen J. Bartels. 2019. Application of community-engaged research to inform the development and implementation of a peer-delivered mobile health intervention for adults with serious mental illness. *Journal of Participatory Medicine* 11(1): e12380.
4. Lisa A. Cooper, Tanjala S. Purnell, Chidinma A. Ibe, Jennifer P. Halbert, Lee R. Bone, Kathryn A. Carson, Debra Hickman, Michelle Simmons, Ann Vachon, Inez Robb, Michelle Martin-Daniels, Katherine B. Dietz, Sherita Hill

Golden, Deidra C. Crews, Felicia Hill-Briggs, Jill A. Marsteller, L. Ebony Boulware, Edgar R. Miller, and David M. Levine. 2016. Reaching for health equity and social justice in Baltimore: The evolution of an academic-community partnership and conceptual framework to address hypertension disparities. *Ethnicity & Disease* 26(3): 369–378.

5. Johns Hopkins Medicine. 2017. Minority health disparities: Michelle's story. *YouTube*, https://www.youtube.com/watch?v=vlVZKZNXYBA.

6. Lisa A. Cooper, Jill A. Marsteller, Gary J. Noronha, Sarah J. Flynn, Kathryn A. Carson, Romsai T. Boonyasai, Cheryl A. Anderson, Hanan J. Aboumatar, Debra L. Roter, Katherine B. Dietz, Edgar R. Miller III, Gregory P. Prokopowicz, Arlene T. Dalcin, Jeanne B. Charleston, Michelle Simmons, and Mary Margaret Huizinga. 2013. A multi-level system quality improvement intervention to reduce racial disparities in hypertension care and control: study protocol. *Implementation Science* 8: 60.

7. Tanvir Hussain, Whitney Franz, Emily Brown, Athena Kan, Mekam Okoye, Katherine Dietz, Kara Taylor, Kathryn A. Carson, Jennifer Halbert, Arlene Dalcin, Cheryl A. M. Anderson, Romsai T. Boonyasai, Michael Albert, Jill A. Marsteller, and Lisa A. Cooper. 2016. The role of care management as a population health intervention to address disparities and control hypertension: A quasi-experimental observational study. *Ethnicity & Disease* 21: 285–294.

8. Jonathan C. Hong, William V. Padula, Ilene L. Hollin, Tanvir Hussain, Katherine B. Dietz, Jennifer P. Dietz, Jill A. Marsteller, and Lisa A. Cooper. 2018. Care management to reduce disparities and control hypertension in primary care: A cost-effectiveness analysis. *Medical Care* 56(2): 179–185.

9. Edgar R. Miller III, Lisa A. Cooper, Kathryn A. Carson, Nae-Yuh Wang, Lawrence J. Appel, Debra Gayles, Jeanne Charleston, Karen White, Na You, Yingjie Weng, Michelle Martin-Daniels, Barbara Bates-Hopkins, Inez Robb, Whitney K. Franz, Emily L. Brown, Jennifer P. Halbert, Michael C. Albert,

Arlene T. Dalcin, and Hsin-Chieh Yeh. 2016. A dietary intervention in urban African Americans: Results of the "Five Plus Nuts and Beans" randomized trial. *American Journal of Preventive Medicine* 50(1):87–95.

10. Lisa A. Cooper, Jill A. Marsteller, Kathryn A. Carson, Katherine B. Dietz, Romsai T. Boonyasai, Carmen Alvarez, Chidinma A. Ibe, Deidra C. Crews, Hsin-Chieh Yeh, Edgar R. Miller III, Cheryl R. Dennison-Himmelfarb, Lisa H. Lubomski, Tanjala S. Purnell, Felicia Hill-Briggs, Nae-Yuh Wang, and RICH LIFE Project Investigators. 2020. The RICH LIFE Project: A cluster randomized pragmatic trial comparing the effectiveness of health system only vs. health system, plus a collaborative/stepped care intervention to reduce hypertension disparities. *American Heart Journal* 226: 94–113.

CHAPTER 5. FROM RESEARCH TO PRACTICE AND POLICY

1. B.N. Greenwood, R.R. Hardeman, L. Huang , and A. Sojourner. 2020. Physician-patient racial concordance and disparities in birthing mortality for newborns. *Proceedings of the National Academy of Sciences* 117(35): 21194–21200.

2. Laura E. Riley. 2015. Ensuring a diverse physician workforce: Progress but more to be done. *JAMA Internal Medicine* 175(10): 1708–1709.

3. Maria Aspan. August 9, 2020. Black women account for less than 3% of U.S. doctors. *Fortune*, https://fortune.com/2020/08/09/health-care-racism-black-women-doctors/.

4. Institute of Medicine of the National Academies. 2008. *Retooling for an Aging America: Building the Health Care Workforce*. Washington, DC: National Academies Press.

5. José Efrain Rodriguez, Kendall M. Campbell, John P. Fogarty, and Roxann Mouratidis. 2014. Underrepresented minority faculty in academic medicine: A systematic review of URM faculty development. *Family Medicine* 46(2): 100–104.

6. Linda Pololi, Lisa A. Cooper, and Phyllis Carr. 2010. Race, disadvantage and faculty experiences in academic medicine. *Journal of General Internal Medicine* 25(12): 1363–1369.
7. Eboni G. Price, Aysegul Gozu, David E. Kern, Neil R. Powe, Gary S. Wand, Sherita Golden, and Lisa A. Cooper. 2005. The role of cultural diversity climate in recruitment, promotion, and retention of faculty in academic medicine. *Journal of General Internal Medicine* 20(7): 565–571; Kara L. Odom, Laura Morgan Roberts, Rachel L. Johnson, and Lisa A. Cooper. 2007. Exploring obstacles to and opportunities for professional success among ethnic minority medical students. *Academic Medicine* 82(2): 146–153; Eboni G. Price, Neil R. Powe, David E. Kern, Sherita Hill Golden, Gary S. Wand, and Lisa A. Cooper. 2009. Improving the diversity climate in academic medicine: Faculty perceptions as a catalyst for institutional change. *Academic Medicine* 84(1): 95–105.
8. Molly Carnes, Patricia G. Devine, Linda Baier Manwell, Angela Byars-Winston, Eve Fine, Cecilia E. Ford, Patrick Forscher, Carol Isaac, Anna Kaatz, Wairimu Magua, Mari Palta, and Jennifer Sheridan. 2015. Effect of an intervention to break the gender bias habit for faculty at one institution: A cluster randomized, controlled trial. *Academic Medicine* 90(2): 221–230.
9. Quinn Capers, Leon McDougle, and Daniel M. Clinchot. 2018. Strategies for achieving diversity through medical school admissions. *Journal of Health Care for the Poor and Underserved* 29(1): 9–18.
10. Sonya G. Smith, Phyllis A. Nsiah-Kumi, Pamela R. Jones, and Rubens J. Pamies. 2009. Pipeline programs in the health professions, part 1: Preserving diversity and reducing health disparities. *Journal of the National Medical Association* 101(9): 836–851; Sonya G. Smith, Phyllis A. Nsiah-Kumi, Pamela R. Jones, and Rubens J. Pamies. 2009. Pipeline programs in the health professions, part 2: The impact of recent legal challenges to affirmative action. *Journal of the National Medical Association* 101(9): 852–863.

11. Javeed Sukhera and Chris Watling. 2018. A framework for integrating implicit bias recognition into health professions education. *Academic Medicine* 93(1):35–40.
12. Lisa A. Cooper, Mary Catherine Beach, and David R. Williams. 2019. Confronting bias and discrimination in health care—When silence is not golden. *JAMA Internal Medicine* 179(12): 1686–1687.
13. Jacqueline K. Nelson, Kevin M. Dunn, and Yin Paradies. 2011. Bystander anti-racism: A review of the literature. *Analyses of Social Issues and Public Policy* 11(1): 262–284.
14. Diana J. Burgess, Mary Catherine Beach, and Somnath Saha. 2017. Mindfulness practice: A promising approach to reducing the effects of clinician implicit bias on patients. *Patient Education and Counseling* 100(2): 372–376; Charles R. Figley, ed. 2002. *Treating Compassion Fatigue*. New York: Brunner/Routledge.
15. Pew Research Center, Internet and Technology. 2019. Mobile phone ownership over time, https://www.pewresearch.org/internet/fact-sheet/mobile/.
16. LaPrincess C. Brewer, Karen L. Fortuna, Clarence Jones, Robert Walker, Sharonne N. Hayes, Christi A. Patten, and Lisa A. Cooper. 2020. Back to the future: Achieving health equity through health informatics and digital health. *JMIR mHealth and uHealth* 8(1): e14512.
17. Jorge A. Rodriguez, Cheryl R. Clark, and David W. Bates. 2020. Digital health equity as a necessity in the 21st Century Cures Act Era. *Journal of the American Medical Association* 323(23): 2381–2382.
18. J. Paul Leigh and Juan Du. October 4, 2018. Effects of minimum wages on population health. Johns Hopkins University, Health Affairs. Health Policy Brief, https://www.healthaffairs.org/do/10.1377/hpb20180622.107025/full/.
19. C. Debra, M. Furr-Holden, Elizabeth D. Nesoff, Victoria Nelson, Adam J. Milam, Mieka Smart, Krim Lacey, Roland J. Thorpe, and Philip J. Leaf. 2018.

Understanding the relationship between alcohol outlet density and life expectancy in Baltimore City: The role of community violence and community disadvantage. *Journal of Community Psychology* 47(1), 63–75.

20. Rachel L. Johnson Thornton, Amelia Greiner, Caroline M. Fichtenberg, Beth J. Feingold, Jonathan M. Ellen, and Jacky M. Jennings. 2013. Achieving a healthy zoning policy in Baltimore: Results of a health impact assessment of the TransForm Baltimore zoning code rewrite. *Public Health Reports* 128 (Suppl. 3):87–103.
21. Paula M. Lantz and Sara Rosenbaum. 2020. The potential and realized impact of the Affordable Care Act on health equity. *Journal of Health Politics, Policy and Law* 45 (5): 831–845.
22. Beth Kowitt, "The One Major Metric Missing from 'Best Hospital' Rankings," *Fortune*, August 16, 2020, https://fortune.com/2020/08/16/best-hospital-rankings-equity/.

CHAPTER 6. A GLOBAL PERSPECTIVE ON HEALTH EQUITY

1. Johns Hopkins Alliance for a Healthier World. ADHINCRA study: Addressing hypertension in Africa, https://www.ahealthierworld.jhu.edu/adhincra-hypertension-in-africa.

CHAPTER 7. HEALTH EQUITY IN THE ERA OF COVID-19

1. Julian Kossoff. April 21, 2020. African Americans seem to be dying of the coronavirus at outsized rates, based on the few states providing data on the race of victims. *Business Insider*, https://www.businessinsider.com/covid-19-disproportionately-kills-african-americans-data-suggests-2020-4.
2. APM Research Lab. 2020. The color of coronavirus: COVID-19 by race and ethnicity in the U.S., https://www.apmresearchlab.org/covid/deaths-by-race.

3. Lisa A. Cooper and Joshua M. Sharfstein. April 7, 2020. A game plan to help the most vulnerable. *Politico,* https://www.politico.com/news/agenda/2020/04/07/game-plan-to-help-those-most-vulnerable-to-covid-19-171863.
4. J. Bor, A.S. Venkataramani, D.R. Williams, and A.C. Tsai. 2018. Police killings and their spillover effects on the mental health of Black Americans: A population-based, quasi-experimental study. *Lancet* 392(10144): 302–310.
5. S.H. Woolf, D.A. Chapman, R.T. Sabo, D.M. Weinberger, and L. Hill. March–July 2020. Excess deaths from COVID-19 and other causes. *Journal of the American Medical Association* 324 (15), 1562–1564.
6. Laurel Wamsley. "American life expectancy dropped by a full year in 1st half of 2020." *NPR*, February 18, 2021, https://www.npr.org/2021/02/18/968791431/american-life-expectancy-dropped-by-a-full-year-in-the-first-half-of-2020.
7. M.R. Jackman and K.A. Shauman. 2019. The toll of inequality: Excess African American deaths in the United States over the twentieth century. *Du Bois Review* 16(2): 291–340.
8. D. Umberson, J.S. Olson, R. Crosnoe, H. Liu, T. Pudrovska, and R. Donnelly. 2017. Death of family members as an overlooked source of racial disadvantage in the United States. *Proceedings of the National Academy of Sciences* 114(5): 915–920.
9. L.A. Cooper and D.R. Williams. October 12, 2020. Excess deaths, community bereavement, and restorative justice for Black and Brown communities. *Journal of the American Medical Association* 324(15): 1491–1492.
10. Lisa A. Cooper and Joshua M. Sharfstein. April 7, 2020. A game plan to help the most vulnerable. *Politico,* https://www.politico.com/news/agenda/2020/04/07/game-plan-to-help-those-most-vulnerable-to-covid-19-171863.

11. David R. Williams and Lisa A. Cooper. 2019. Reducing racial inequities in health: using what we already know to take action. *International Journal of Environmental Research and Public Health* 16:606.
12. Lisa A. Cooper. 2020. The silver lining in COVID-19's dark clouds. State of Black America, https://soba.iamempowered.com/silver-lining-covid-19%E2%80%99s-dark-clouds.
13. National Institutes of Health, news release. September 16, 2020. NIH funds community engagement research efforts in areas hardest hit by COVID-19, https://www.nih.gov/news-events/news-releases/nih-funds-community-engagement-research-efforts-areas-hardest-hit-covid-19.
14. Joseph P. Williams. March 25, 2020. Rumor, disparity, and distrust: Why Black Americans face an uphill battle against COVID-19. *U.S. News & World Report,* https://www.usnews.com/news/healthiest-communities/articles/2020-03-25/why-black-americans-face-an-uphill-battle-against-the-coronavirus.
15. Jennifer K. Benz, Oscar Espinosa, Valerie Welsh, and Angela Fontes. 2011. Awareness of racial and ethnic health disparities has improved only modestly over a decade. *Health Affairs* 30(1): 1860–1867.
16. Jessica Kendrick, Eugene Nuccio, Jenn A. Leiferman, and Angela Sauaia. 2015. Primary care providers perceptions of racial/ethnic and socioeconomic disparities in hypertension control. *American Journal of Hypertension* 28(9): 1091–1097.
17. David R. Williams and Lisa A. Cooper. 2020. COVID-19 and health equity—a new kind of "herd immunity." *Journal of the American Medical Association* 323(24): 2478–2480.
18. Justin Frewen and Anna Datta. 2010. The cost of mental health in Europe, https://www.worldpress.org/Europe/3613.cfm.
19. Richard G. Wilkinson. 2005. *The Impact of Inequality: How to Make Sick Societies Healthier.* New York: Routledge.

Index